Farm Management

Farm Management

Edited by
Wilfrid Stevens

Larsen & Keller
www.larsen-keller.com

Farm Management
Edited by Wilfrid Stevens
ISBN: 978-1-63549-118-0 (Hardback)

☰ Larsen & Keller

Published by Larsen and Keller Education,
5 Penn Plaza,
19th Floor,
New York, NY 10001, USA

Cataloging-in-Publication Data

Farm management / edited by Wilfrid Stevens.
    p. cm.
Includes bibliographical references and index.
ISBN 978-1-63549-118-0
1. Farm management. 2. Agricultural productivity.
3. Soils and nutrition. 4. Plants--Nutrition. 5. Agriculture.
I. Stevens, Wilfrid.
S561 .F37 2017
630--dc23

The publisher's policy is to use permanent paper from mills that operate a sustainable forestry policy. Furthermore, the publisher ensures that the text paper and cover boards used have met acceptable environmental accreditation standards.

Printed and bound in the United States of America.

For more information regarding Larsen and Keller Education and its products, please visit the publisher's website www.larsen-keller.com

Table of Contents

Preface

Farm management is an interdisciplinary field of agricultural sciences. It includes many different topics like livestock management, agronomy, biology of farm animals, husbandry, etc. This book provides various concepts and applications of the subject. It describes the fundamentals of this subject area in detail. The text picks up individual methods and explains their need and contribution in the context of the growth of this field. The topics included in this book on farm management are of utmost significance and bound to provide incredible insights to readers. Coherent flow of topics, student-friendly language and extensive use of examples make this book an invaluable source of knowledge. It will be of great help to students in the fields of daily science, livestock management and animal breeding.

A detailed account of the significant topics covered in this book is provided below:

Chapter 1- Farm is an area that is used for agricultural practices. The purpose of having livestock is to produce food and crops; it is mostly used for food production. This section is an overview of the subject matter incorporating all the major aspects of farms.

Chapter 2- Agricultural science is a branch of biology that encompasses the procedures and practices involved in agriculture. Agricultural science is not the same as agronomy or agriculture, all three are different concepts. Agricultural science includes production techniques, minimizing the effects of pests, improving agricultural productivity and the prevention of adverse environmental effects. The following text will provide an insightful focus, keeping in mind the subject matter.

Chapter 3- Dairy farming is done with the motive of obtaining long-term production of milk. The animals used in dairy farming include goats, sheeps and camels. Some of the topics discussed in this section are dairy cattle, milking pipeline, automatic milking, bovine somatotropin, dairy product and bulk tank. This chapter will provide an integrated understanding of dairy farming.

Chapter 4- Raising domesticated birds is known as poultry farming. These birds include ducks, chickens and geese and are used for the purpose of farming eggs and meat. Poultry, free range, yarding, battery cage, furnished cages and broiler industry are some of the aspects of poultry farming. The aspects elucidated in this section help the reader in developing a better understanding of poultry farming.

Chapter 5- Organic farming consciously promotes agricultural practices that do not harm the environment. Organic farming mainly relies on organic fertilizers such as manure, green manure and bone meal. Some of the important aspects dealt within this section are organic horticulture, natural farming, organic certification and organic fertilizer. The topics elaborated in this text will help in gaining a better perspective on organic farming.

Chapter 6- Agricultural productivity is the ratio of agriculture outputs to agricultural inputs. The major processes of agricultural productivity elucidated in this chapter are mechanised agriculture, irrigation, theoretical production ecology, leaf area index and animal feed. This chapter has been carefully written to provide an easy understanding of the varied processes of agricultural productivity.

Chapter 7- The ability of soil to nurture the growth of a plant is known as soil fertility. Soil conversation is the prevention of soil loss from erosion. The methods of soil conservation include crop rotation, cover crop, tillage and windbreak. In order to completely understand soil nutrition and soil sampling, it is necessary to understand the processes related to it. The following section elucidates the processes associated with soil nutrition.

Chapter 8- Plant nutrition is the analysis of chemical elements that are vital for plant growth and plant metabolism. For plant nutrition, water potential and transpiration are also important processes involved in the growth of a plant. This text helps the reader in understanding the concept of plant nutrition.

I would like to make a special mention of my publisher who considered me worthy of this opportunity and also supported me throughout the process. I would also like to thank the editing team at the back-end who extended their help whenever required.

Editor

Introduction to Farm

Farm is an area that is used for agricultural practices. The purpose of having livestock is to produce food and crops; it is mostly used for food production. This section is an overview of the subject matter incorporating all the major aspects of farms.

A farm is an area of land that is devoted primarily to agricultural processes with the primary objective of producing food and other crops; it is the basic facility in food production. The name is used for specialised units such as arable farms, vegetable farms, fruit farms, dairy, pig and poultry farms, and land used for the production of natural fibres, biofuel and other commodities. It includes ranches, feedlots, orchards, plantations and estates, smallholdings and hobby farms, and includes the farmhouse and agricultural buildings as well as the land. In modern times the term has been extended so as to include such industrial operations as wind farms and fish farms, both of which can operate on land or sea.

Farmland in the USA. The round fields are due to the use of center pivot irrigation

Farming originated independently in different parts of the world as hunter gatherer societies transitioned to food production rather than food capture. It may have started about 12,000 years ago with the domestication of livestock in the Fertile Crescent in western Asia, soon to be followed by the cultivation of crops. Modern units tend to specialise in the crops or livestock best suited to the region, with their finished products being sold for the retail market or for further processing, with farm products being traded around the world.

Modern farms in developed countries are highly mechanized. In the United States, livestock may be raised on rangeland and finished in feedlots and the mechanisation of crop production has brought about a great decrease in the number of agricultural workers needed. In Europe, traditional family farms are giving way to larger production units. In Australia, some farms are very large because the land is unable to support a high stocking density of livestock because of climatic conditions. In less developed countries, small farms are the norm, and the majority of rural residents are subsistence farmers, feeding their families and selling any surplus products in the local market.

Typical plan of a mediaeval English manor, showing the use of field strips

Etymology

A farmer harvesting crops with mule-drawn wagon, 1920s, Iowa, USA

The word in the sense of an agricultural land-holding derives from the verb "to farm" a revenue source, whether taxes, customs, rents of a group of manors or simply to hold an individual manor by the feudal land tenure of "fee farm". The word is from the medieval Latin noun *firma*, also the source of the French word *ferme*, meaning a fixed agreement, contract, from the classical Latin adjective *firmus* meaning strong, stout, firm. As in the medieval age virtually all manors were engaged in the business of agriculture, which was their principal revenue source, so to hold a manor by the tenure of "fee farm" became synonymous with the practice of agriculture itself.

History

Farming has been innovated at multiple different points and places in human history. The transition from hunter-gatherer to settled, agricultural societies is called the Neolithic Revolution and first began around 12,000 years ago, near the beginning of the geological epoch of the Holocene around 12,000 years ago. It was the world's first historically verifiable revolution in agriculture.

Subsequent step-changes in human farming practices were provoked by the British Agricultural Revolution in the 18th century, and the Green Revolution of the second half of the 20th century. Farming spread from the Middle East to Europe and by 4,000 BC people that lived in the central part of Europe were using oxen to pull plows and wagons.

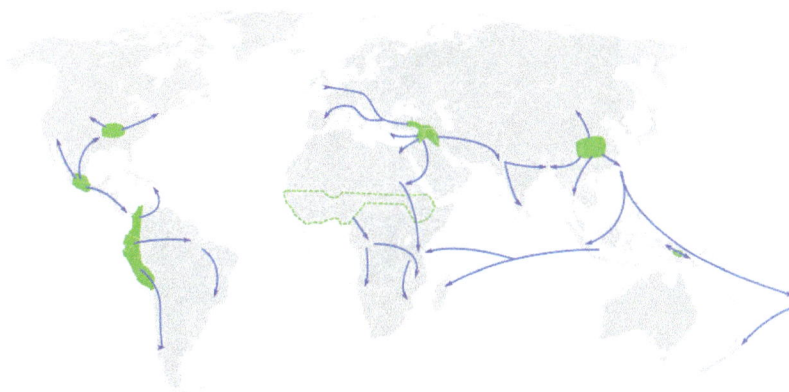

Map of the world showing approximate centers of origin of agriculture and its spread in prehistory: the Fertile Crescent (11,000 BP), the Yangtze and Yellow River basins (9,000 BP), and the New Guinea Highlands (9,000–6,000 BP), Central Mexico (5,000–4,000 BP), Northern South America (5,000–4,000 BP), sub-Saharan Africa (5,000–4,000 BP, exact location unknown), eastern North America (4,000–3,000 BP).

Types of Farm

A farm may be owned and operated by a single individual, family, community, corporation or a company, may produce one or many types of produce, and can be a holding of any size from a fraction of a hectare to several thousand hectares.

A farm may operate under a monoculture system or with a variety of cereal or arable crops, which may be separate from or combined with raising livestock. Specialist farms are often denoted as such, thus a dairy farm, fish farm, poultry farm or mink farm.

Some farms may not use the word at all, hence vineyard (grapes), orchard (nuts and other fruit), market garden or "truck farm" (vegetables and flowers). Some farms may be denoted by their topographical location, such as a hill farm, while large estates growing cash crops such as cotton or coffee may be called plantations.

Many other terms are used to describe farms to denote their methods of production, as in collective, corporate, intensive, organic or vertical.

Other farms may primarily exist for research or education, such as an ant farm, and since farming is synonymous with mass production, the word "farm" may be used to describe wind power generation or puppy farm.

Specialized Farms

Dairy Farm

Dairy farming is a class of agriculture, where female cattle, goats, or other mammals are raised for their milk, which may be either processed on-site or transported to a dairy for processing and eventual retail sale. There are many breeds of cattle that can be milked some of the best producing ones include Holstein, Norwegian Red, Kostroma, Brown Swiss, and more.

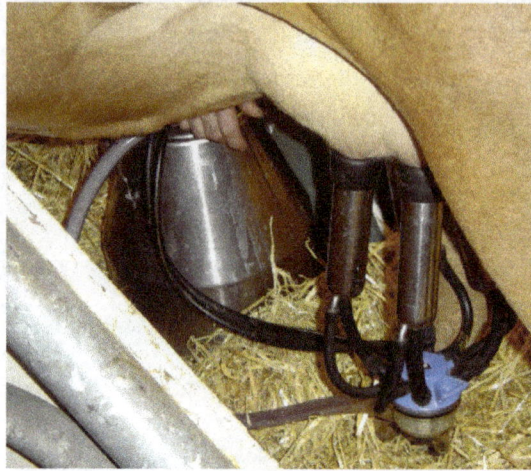

A milking machine in action

In most Western countries, a centralized dairy facility processes milk and dairy products, such as cream, butter, and cheese. In the United States, these dairies are usually local companies, while in the southern hemisphere facilities may be run by very large nationwide or trans-national corporations (such as Fonterra).

Dairy farms generally sell male calves for veal meat, as dairy breeds are not normally satisfactory for commercial beef production. Many dairy farms also grow their own feed, typically including corn, alfalfa, and hay. This is fed directly to the cows, or stored as silage for use during the winter season. Additional dietary supplements are added to the feed to improve milk production.

Poultry Farm

Poultry farming

Poultry farms are devoted to raising chickens (egg layers or broilers), turkeys, ducks, and other fowl, generally for meat or eggs.

Pig Farm

A pig farm is one that specializes in raising pigs or hogs for bacon, ham and other pork products and may be free range, intensive, or both.

Prison Farm

Prison farms are farms which serve as prisons for people sentenced to hard labor by a court. On prison farms inmates run the important tasks of a farm and producing crops.

Ownership

Farm control and ownership has traditionally been a key indicator of status and power, especially in Medieval European agrarian societies. The distribution of farm ownership has historically been closely linked to form of government. Medieval feudalism was essentially a system that centralized control of farmland, control of farm labor and political power, while the early American democracy, in which land ownership was a prerequisite for voting rights, was built on relatively easy paths to individual farm ownership. However, the gradual modernization and mechanization of farming, which greatly increases both the efficiency and capital requirements of farming, has led to increasingly large farms. This has usually been accompanied by the decoupling of political power from farm ownership.

Forms of Ownership

In some societies (especially socialist and communist), collective farming is the norm, with either government ownership of the land or common ownership by a local group. Especially in societies without widespread industrialized farming, tenant farming and sharecropping are common; farmers either pay landowners for the right to use farmland or give up a portion of the crops.

Farms around the World

Americas

Farming near Klingerstown, Pennsylvania

The land and buildings of a farm are called the "farmstead". Enterprises where livestock are raised on rangeland are called *ranches*. Where livestock are raised in confinement on feed produced elsewhere, the term *feedlot* is usually used.

A typical North American grain farm with farmstead in Ontario, Canada

In 1910 there were 6,406,000 farms and 10,174,000 family workers; In 2000 there were only 2,172,000 farms and 2,062,300 family workers. The share of U.S. farms operated by women has risen steadily over recent decades, from 5 percent in 1978 to 14 percent by 2007.

In the United States, there are over three million migrant and seasonal farmworkers; 72% are foreign-born, 78% are male, they have an average age of 36 and average education of 8 years. Farmworkers make an average hourly rate of $9–10 per hour, compared to an average of over $18 per hour for nonfarm labor. Their average family income is under $20,000 and 23% live in families with incomes below the federal poverty level. One-half of all farmworker families earn less than $10,000 per year, which is significantly below the 2005 U.S. poverty level of $19,874 for a family of four.

In 2007, corn acres are expected to increase by 15% because of the high demand for ethanol, both in and outside of the U.S. Producers are expecting to plant 90.5 million acres (366,000 km^2) of corn, making it the largest corn crop since 1944.

Asia

Farmlands in Hebei province, China

Pakistan

According to the World Bank, "most empirical evidence indicates that land productivity on large farms in Pakistan is lower than that of small farms, holding other factors constant." Small farmers have "higher net returns per hectare" than large farms, according to farm household income data.

Nepal

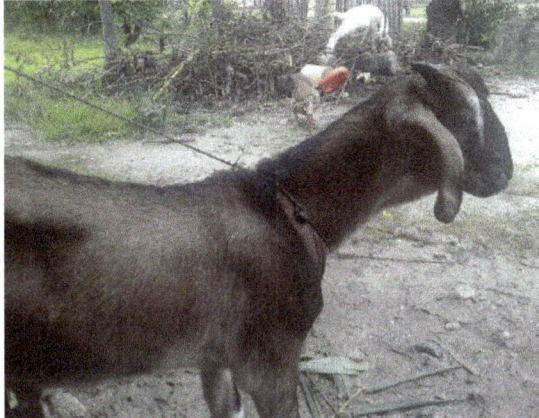
Goat found in Nepal

Nepal is an agricultural country and about 80% of the total population are engaged in farming. Rice is mainly produced in Nepal along with fruits like apples. Dairy farming and poultry farming are also growing in Nepal.

Australia

Cows grazing on a farm in Victoria, Australia

Farming is a significant economic sector in Australia. A farm is an area of land used for primary production which will include buildings.

According to the UN, "green agriculture directs a greater share of total farming input expenditures towards the purchase of locally sourced inputs (e.g. labour and organic fertilisers) and a local multiplier effect is expected to kick in. Overall, green farming practices tend to require more labour inputs than conventional farming (e.g. from comparable levels to as much as 30 per cent more) (FAO 2007 and European Commission 2010), creating jobs in rural areas and a higher return on labour inputs."

Where most of the income is from some other employment, and the farm is really an expanded residence, the term *hobby farm* is common. This will allow sufficient size for recreational use but be very unlikely to produce sufficient income to be self-sustaining. Hobby farms are commonly around 2 hectares (4.9 acres) but may be much larger depending upon land prices (which vary regionally).

Often very small farms used for intensive primary production are referred to by the specialization they are being used for, such as a dairy rather than a dairy farm, a piggery, a market garden, etc. This also applies to feedlots, which are specifically developed to a single purpose and are often not able to be used for more general purpose (mixed) farming practices.

In remote areas farms can become quite large. As with *estates* in England, there is no defined size or method of operation at which a large farm becomes a station.

Europe

Traditional Dutch farmhouse

In the UK, *farm* as an agricultural unit, always denotes the area of pasture and other fields together with its farmhouse, farmyard and outbuildings. Large farms, or groups of farms under the same ownership, may be called an estate. Conversely, a small farm surrounding the owner's dwelling is called a smallholding and is generally focused on self-sufficiency with only the surplus being sold.

Farm Equipment

Farm equipment has evolved over the centuries from simple hand tools such as the hoe, through ox- or horse-drawn equipment such as the plough and harrow, to the modern highly-technical machinery such as the tractor, baler and combine harvester replacing what was a highly labour-intensive occupation before the Industrial revolution. Today much of the farm equipment used on both small and large farms is automated (e.g. using satellite guided farming).

As new types of high-tech farm equipment have become inaccessible to farmers that historically fixed their own equipment, Wired reports there is a growing backlash, due mostly to companies using intellectual property law to prevent farmers from having the legal right to fix their equipment (or gain access to the information to allow them to do it). This has encouraged groups such as Open Source Ecology and Farm Hack to begin to make open source hardware for agricultural machinery. In addition on a smaller scale Farmbot and the RepRap open source 3D printer community has begun to make open-source farm tools available of increasing levels of sophistication.

References

- Graeme Barker (25 March 2009). The Agricultural Revolution in Prehistory: Why did Foragers become Farmers?. Oxford University Press. ISBN 978-0-19-955995-4. Retrieved 15 August 2012.

- Hoppe, Robert A. and Penni Korb. (2013). Characteristics of Women Farm Operators and Their Farms. Washington, D.C.: U.S. Department of Agriculture, Economic Research Service.

- "RSS Text Size Print Share This Home / news / opinion / editorial / Taxpayers Get a Break From Prison Farms". The News & Advance. August 28, 2008. Retrieved February 18, 2012.

- "Corn Acres Expected to Soar in 2007, USDA Says". Newsroom. Washington: U.S. Department of Agriculture - National Agricultural Statistics Service. March 30, 2007. Retrieved February 18, 2012.

- http://www.slate.com/articles/technology/future_tense/2012/06/automated_farm_equipment_and_small_scale_farmers_.html

- http://www.unep.org/greeneconomy/Portals/88/documents/ger/ger_final_dec_2011/Green%20EconomyReport_Final_Dec2011.pdf

Understanding Farm Management

Agricultural science is a branch of biology that encompasses the procedures and practices involved in agriculture. Agricultural science is not the same as agronomy or agriculture, all three are different concepts. Agricultural science includes production techniques, minimizing the effects of pests, improving agricultural productivity and the prevention of adverse environmental effects. The following text will provide an insightful focus, keeping in mind the subject matter.

Agricultural science is a broad multidisciplinary field of biology that encompasses the parts of exact, natural, economic and social sciences that are used in the practice and understanding of agriculture. (Veterinary science, but not animal science, is often excluded from the definition.)

Agriculture, Agricultural Science, and Agronomy

The three terms are often confused. However, they cover different concepts:

- Agriculture is the set of activities that transform the environment for the production of animals and plants for human use. Agriculture concerns techniques, including the application of agronomic research.

- Agronomy is research and development related to studying and improving plant-based crops.

Agricultural sciences include research and development on:

- Production techniques (e.g., irrigation management, recommended nitrogen inputs)

- Improving agricultural productivity in terms of quantity and quality (e.g., selection of drought-resistant crops and animals, development of new pesticides, yield-sensing technologies, simulation models of crop growth, in-vitro cell culture techniques)

- Minimizing the effects of pests (weeds, insects, pathogens, nematodes) on crop or animal production systems.

- Transformation of primary products into end-consumer products (e.g., production, preservation, and packaging of dairy products)

- Prevention and correction of adverse environmental effects (e.g., soil degradation, waste management, bioremediation)

- Theoretical production ecology, relating to crop production modeling

- Traditional agricultural systems, sometimes termed subsistence agriculture, which feed most of the poorest people in the world. These systems are of interest as they sometimes retain a level of integration with natural ecological systems greater than that of industrial agriculture, which may be more sustainable than some modern agricultural systems.

- Food production and demand on a global basis, with special attention paid to the major producers, such as China, India, Brazil, the USA and the EU.

- Various sciences relating to agricultural resources and the environment (e.g. soil science, agroclimatology); biology of agricultural crops and animals (e.g. crop science, animal science and their included sciences, e.g. ruminant nutrition, farm animal welfare); such fields as agricultural economics and rural sociology; various disciplines encompassed in agricultural engineering.

Agricultural Biotechnology

Agricultural biotechnology is a specific area of agricultural science involving the use of scientific tools and techniques, including genetic engineering, molecular markers, molecular diagnostics, vaccines, and tissue culture, to modify living organisms: plants, animals, and microorganisms.

Fertilizer

One of the most common yield reducers is because of fertilizer not being applied in slightly higher quantities during transition period, the time it takes the soil to rebuild its aggregates and organic matter. Yields will decrease temporarily because of nitrogen being immobilized in the crop residue, which can take a few months to several years to decompose, depending on the crop's C to N ratio and the local environment

A local Science

With the exception of theoretical agronomy, research in agronomy, more than in any other field, is strongly related to local areas. It can be considered a science of ecoregions, because it is closely linked to soil properties and climate, which are never exactly the same from one place to another. Many people think an agricultural production system relying on local weather, soil characteristics, and specific crops has to be studied locally. Others feel a need to know and understand production systems in as many areas as possible, and the human dimension of interaction with nature.

History of Agricultural Science

Agricultural science began with Gregor Mendel's genetic work, but in modern terms might be better dated from the chemical fertilizer outputs of plant physiological understanding in 18th-century Germany. In the United States, a scientific revolution in agriculture began with the Hatch Act of 1887, which used the term "agricultural science". The Hatch Act was driven by farmers' interest in knowing the constituents of early artificial fertilizer. The Smith-Hughes Act of 1917 shifted agricultural education back to its vocational roots, but the scientific foundation had been built. After 1906, public expenditures on agricultural research in the US exceeded private expenditures for the next 44 years.

Intensification of agriculture since the 1960s in developed and developing countries, often referred to as the Green Revolution, was closely tied to progress made in selecting and improving crops and animals for high productivity, as well as to developing additional inputs such as artificial fertilizers and phytosanitary products.

As the oldest and largest human intervention in nature, the environmental impact of agriculture in general and more recently intensive agriculture, industrial development, and population growth have raised many questions among agricultural scientists and have led to the development and emergence of new fields. These include technological fields that assume the solution to technological problems lies in better technology, such as integrated pest management, waste treatment technologies, landscape architecture, genomics, and agricultural philosophy fields that include references to food production as something essentially different from non-essential economic 'goods'. In fact, the interaction between these two approaches provide a fertile field for deeper understanding in agricultural science.

New technologies, such as biotechnology and computer science (for data processing and storage), and technological advances have made it possible to develop new research fields, including genetic engineering, agrophysics, improved statistical analysis, and precision farming. Balancing these, as above, are the natural and human sciences of agricultural science that seek to understand the human-nature interactions of traditional agriculture, including interaction of religion and agriculture, and the non-material components of agricultural production systems.

Prominent Agricultural Scientists

- Robert Bakewell
- Norman Borlaug
- Luther Burbank
- George Washington Carver
- René Dumont
- Sir Albert Howard
- Kailas Nath Kaul
- Justus von Liebig
- Jay Lush
- Gregor Mendel
- Louis Pasteur
- M. S. Swaminathan
- Jethro Tull
- Artturi Ilmari Virtanen
- Eli Whitney
- Sewall Wright

Agricultural Science and Agriculture Crisis

Agriculture sciences seek to feed the world's population while preventing biosafety problems that may affect human health and the environment. This requires promoting good management of natural resources and respect for the environment, and increasingly concern for the psychological wellbeing of all concerned in the food production and consumption system.

Economic, environmental, and social aspects of agriculture sciences are subjects of ongoing debate. Recent crises (such as avian influenza, mad cow disease and issues such as the use of genetically modified organisms) illustrate the complexity and importance of this debate.

Fields or Related Disciplines

- Agricultural biotechnology
- Agricultural chemistry
- Agricultural diversification
- Agricultural education
- Agricultural economics
- Agricultural engineering
- Agricultural geography
- Agricultural philosophy
- Agricultural marketing
- Agricultural soil science
- Agroecology
- Agrophysics
- Animal science
    - Animal breeding
    - Animal husbandry
    - Animal nutrition
- Agronomy
    - Botany
    - Theoretical production ecology
    - Horticulture
    - Plant breeding
    - Plant fertilization

- Aquaculture
- Biological engineering
    - Genetic engineering
- Nematology
- Microbiology
    - Plant pathology
- Range management
- Environmental science
- Entomology
- Food science
    - Human nutrition
- Irrigation and water management
- Soil science
    - Agrology
- Waste management
- Weed science

Dairy Farming: An Integrated Study

Dairy farming is done with the motive of obtaining long-term production of milk. The animals used in dairy farming include goats, sheeps and camels. Some of the topics discussed in this section are dairy cattle, milking pipeline, automatic milking, bovine somatotropin, dairy product and bulk tank. This chapter will provide an integrated understanding of dairy farming.

Dairy Farming

Dairy farming is a class of agriculture for long-term production of milk, which is processed (either on the farm or at a dairy plant, either of which may be called a dairy) for eventual sale of a dairy product.

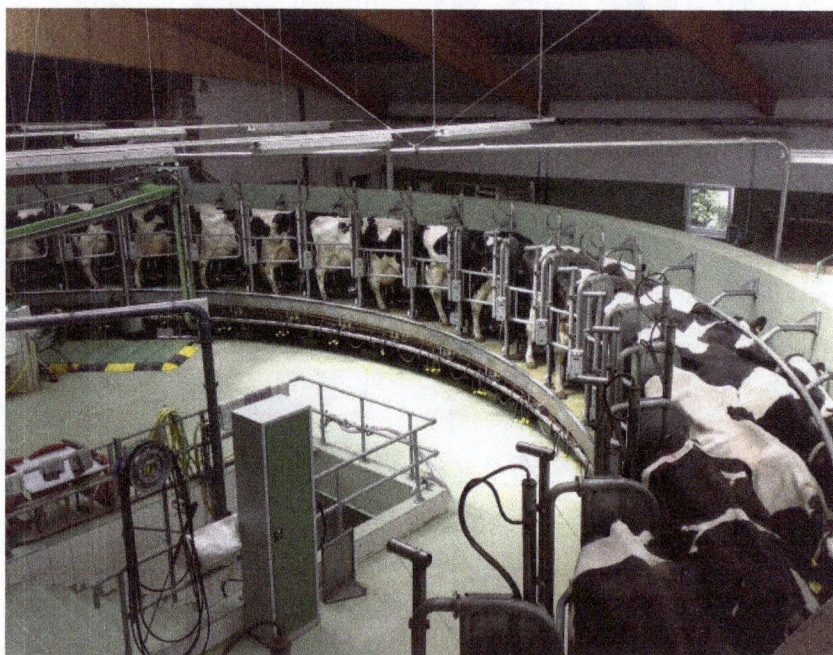

A rotary milking parlor at a modern dairy facility, located in Germany

Common Species

Although any mammal can produce milk, commercial dairy farms are typically one-species enterprises. In developed countries, dairy farms typically consist of high producing dairy cows. Other species used in commercial dairy farming include goats, sheep, and camels. In Italy, donkey dairies are growing in popularity to produce an alternative milk source for human infants.

A dairy farm on the banks of the Columbia River in Clark County, Washington (May 1973).

History

Dairy farming has been part of agriculture for thousands of years. Historically it has been one part of small, diverse farms. In the last century or so larger farms doing only dairy production have emerged. Large scale dairy farming is only viable where either a large amount of milk is required for production of more durable dairy products such as cheese, butter, etc. or there is a substantial market of people with cash to buy milk, but no cows of their own.

Hand Milking

Woman hand milking a cow.

Centralized dairy farming as we understand it primarily developed around villages and cities, where residents were unable to have cows of their own due to a lack of grazing land. Near the town, farmers could make some extra money on the side by having additional animals and selling the milk in town. The dairy farmers would fill barrels with milk in the morning and bring it to market on a wagon. Until the late 19th century, the milking of the cow was done by hand. In the United States, several large dairy operations existed in some northeastern states and in the west, that involved as many as several hundred cows, but an individual milker could not be expected to milk more than a dozen cows a day. Smaller operations predominated.

For most herds, milking took place indoors twice a day, in a barn with the cattle tied by the neck with ropes or held in place by stanchions. Feeding could occur simultaneously with milking in the barn, although most dairy cattle were pastured during the day between milkings. Such examples of this method of dairy farming are difficult to locate, but some are preserved as a historic site for a glimpse into the days gone by. One such instance that is open for this is at Point Reyes National Seashore.

Dairy farming has been part of agriculture for thousands of years. Historically it has been one part of small, diverse farms. In the last century or so larger farms doing only dairy production have emerged. Large scale dairy farming is only viable where either a large amount of milk is required for production of more durable dairy products such as cheese, butter, etc. or there is a substantial market of people with cash to buy milk, but no cows of their own.

Vacuum Bucket Milking

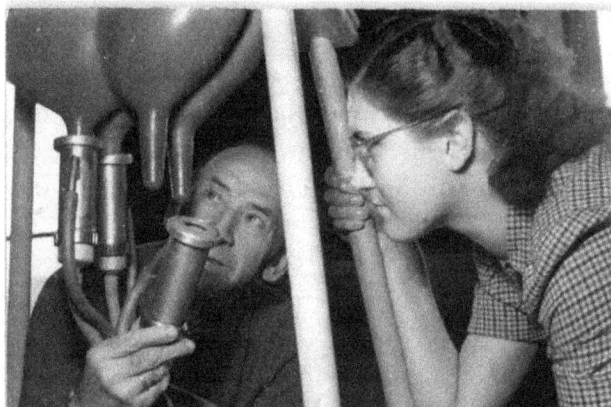

Demonstration of a new Soviet milker device. East Germany, 1952

The first milking machines were an extension of the traditional milking pail. The early milker device fit on top of a regular milk pail and sat on the floor under the cow. Following each cow being milked, the bucket would be dumped into a holding tank. These were introduced in the early 20th century.

This developed into the Surge hanging milker. Prior to milking a cow, a large wide leather strap called a surcingle was put around the cow, across the cow's lower back. The milker device and collection tank hung underneath the cow from the strap. This innovation allowed the cow to move around naturally during the milking process rather than having to stand perfectly still over a bucket on the floor.

Milking Pipeline

The next innovation in automatic milking was the milk pipeline, introduced in the late 20th century. This uses a permanent milk-return pipe and a second vacuum pipe that encircles the barn or milking parlor above the rows of cows, with quick-seal entry ports above each cow. By eliminating the need for the milk container, the milking device shrank in size and weight to the point where it could hang under the cow, held up only by the sucking force of the milker nipples on the cow's udder. The milk is pulled up into the milk-return pipe by the vacuum system, and then flows by gravity to the milkhouse vacuum-breaker that puts the milk in the storage tank. The pipeline system greatly reduced the physical labor of milking since the farmer no longer needed to carry around huge heavy buckets of milk from each cow.

The pipeline allowed barn length to keep increasing and expanding, but after a point farmers started to milk the cows in large groups, filling the barn with one-half to one-third of the herd, milking the animals, and then emptying and refilling the barn. As herd sizes continued to increase, this evolved into the more efficient milking parlor.

Milking Parlors

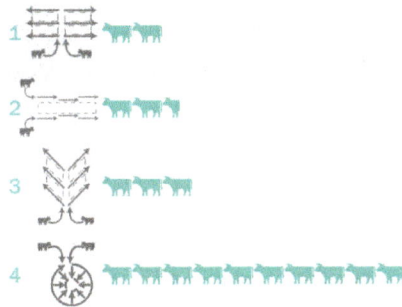

Efficiency of four different milking parlors.
1=Bali-Style 50 cows/h.
2=Swingover 60 cows/h.
3=Herringbone 75 cows/h.
4=Rotary 250 cows/h.

Innovation in milking focused on mechanizing the milking parlor (known in Australia and New Zealand as a *milking shed*) to maximize the number of cows per operator which streamlined the milking process to permit cows to be milked as if on an assembly line, and to reduce physical stresses on the farmer by putting the cows on a platform slightly above the person milking the cows to eliminate having to constantly bend over. Many older and smaller farms still have tie-stall or stanchion barns, but worldwide a majority of commercial farms have parlors.

Herringbone and Parallel Parlors

In herringbone and parallel parlors, the milker generally milks one row at a time. The milker will move a row of cows from the holding yard into the milking parlor, and milk each cow in that row. Once all of the milking machines have been removed from the milked row, the milker releases the cows to their feed. A new group of cows is then loaded into the now vacant side and the process repeats until all cows are milked. Depending on the size of the milking parlor, which normally is the bottleneck, these rows of cows can range from four to sixty at a time.

Rotary Parlors

In rotary parlors, the cows are loaded one at a time onto the platform as it rotates. The milker stands near the entry to the parlor and puts the cups on the cows as they move past. By the time the platform has completed almost a full rotation, another milker or a machine removes the cups and the cow steps backwards off the platform and then walks to its feed. Rotary cowsheds, as they are called in New Zealand, started in the 1980s but are expensive compared to Herringbone cowshed - the older New Zealand norm.

Rotary milking parlor

Automatic Milker Take-off

It can be harmful to an animal for it to be over-milked past the point where the udder has stopped releasing milk. Consequently, the milking process involves not just applying the milker, but also monitoring the process to determine when the animal has been *milked out* and the milker should be removed. While parlor operations allowed a farmer to milk many more animals much more quickly, it also increased the number of animals to be monitored simultaneously by the farmer. The automatic take-off system was developed to remove the milker from the cow when the milk flow reaches a preset level, relieving the farmer of the duties of carefully watching over 20 or more animals being milked at the same time.

Fully Automated Robotic Milking

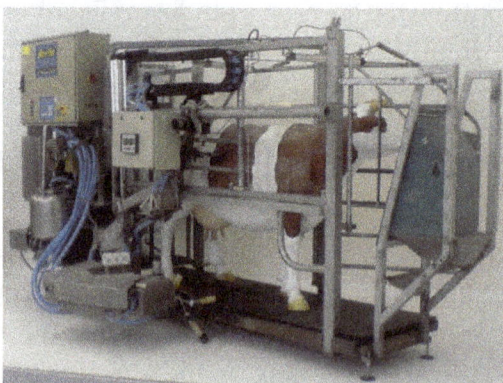
An automatic milking system unit as an exhibit at a museum

In the 1980s and 1990s, robotic milking systems were developed and introduced (principally in the EU). Thousands of these systems are now in routine operation. In these systems the cow has a high degree of autonomy to choose her time of milking freely during the day (some alternatives may apply, depending on cow-traffic solution used at a farm level). These systems are generally limited to intensively managed systems although research continues to match them to the requirements of grazing cattle and to develop sensors to detect animal health and fertility automatically. Every time the cow enters the milking unit she is fed concentrates and her collar is scanned to record production data.

History of Milk Preservation Methods

Cool temperature has been the main method by which milk freshness has been extended. When windmills and well pumps were invented, one of their first uses on the farm, besides providing water for animals themselves, was for cooling milk, to extend its storage life, until it would be transported to the town market.

The naturally cold underground water would be continuously pumped into a cooling tub or vat. Tall, ten-gallon metal containers filled with freshly obtained milk, which is naturally warm, were placed in this cooling bath. This method of milk cooling was popular before the arrival of electricity and refrigeration.

Refrigeration

When refrigeration first arrived (the 19th century) the equipment was initially used to cool cans of milk, which were filled by hand milking. These cans were placed into a cooled water bath to remove heat and keep them cool until they were able to be transported to a collection facility. As more automated methods were developed for harvesting milk, hand milking was replaced and, as a result, the milk can was replaced by a bulk milk cooler. 'Ice banks' were the first type of bulk milk cooler. This was a double wall vessel with evaporator coils and water located between the walls at the bottom and sides of the tank. A small refrigeration compressor was used to remove heat from the evaporator coils. Ice eventually builds up around the coils, until it reaches a thickness of about three inches surrounding each pipe, and the cooling system shuts off. When the milking operation starts, only the milk agitator and the water circulation pump, which flows water across the ice and the steel walls of the tank, are needed to reduce the incoming milk to a temperature below 5 degrees.

This cooling method worked well for smaller dairies, however was fairly inefficient and was unable to meet the increasingly higher cooling demand of larger milking parlors. In the mid-1950s direct expansion refrigeration was first applied directly to the bulk milk cooler. This type of cooling utilizes an evaporator built directly into the inner wall of the storage tank to remove heat from the milk. Direct expansion is able to cool milk at a much faster rate than early ice bank type coolers and is still the primary method for bulk tank cooling today on small to medium-sized operations.

Another device which has contributed significantly to milk quality is the plate heat exchanger (PHE). This device utilizes a number of specially designed stainless steel plates with small spaces between them. Milk is passed between every other set of plates with water being passed between the balance of the plates to remove heat from the milk. This method of cooling can remove large amounts of heat from the milk in a very short time, thus drastically slowing bacteria growth and thereby improving milk quality. Ground water is the most common source of cooling medium for this device. Dairy cows consume approximately 3 gallons of water for every gallon of milk production and prefer to drink slightly warm water as opposed to cold ground water. For this reason, PHE's can result in drastically improved milk quality, reduced operating costs for the dairymen by reducing the refrigeration load on his bulk milk cooler, and increased milk production by supplying the cows with a source of fresh warm water.

Plate heat exchangers have also evolved as a result of the increase of dairy farm herd sizes in the

United States. As a dairyman increases the size of his herd, he must also increase the capacity of his milking parlor in order to harvest the additional milk. This increase in parlor sizes has resulted in tremendous increases in milk throughput and cooling demand. Today's larger farms produce milk at a rate which direct expansion refrigeration systems on bulk milk coolers cannot cool in a timely manner. PHE's are typically utilized in this instance to rapidly cool the milk to the desired temperature (or close to it) before it reaches the bulk milk tank. Typically, ground water is still utilized to provide some initial cooling to bring the milk to between 55 and 70 °F (21 °C). A second (and sometimes third) section of the PHE is added to remove the remaining heat with a mixture of chilled pure water and propylene glycol. These chiller systems can be made to incorporate large evaporator surface areas and high chilled water flow rates to cool high flow rates of milk.

Milking Operation

Milking machines are held in place automatically by a vacuum system that draws the ambient air pressure down from 15 to 21 pounds per square inch (100 to 140 kPa) of vacuum. The vacuum is also used to lift milk vertically through small diameter hoses, into the receiving can. A milk lift pump draws the milk from the receiving can through large diameter stainless steel piping, through the plate cooler, then into a refrigerated bulk tank.

Milk is extracted from the cow's udder by flexible rubber sheaths known as liners or inflations that are surrounded by a rigid air chamber. A pulsating flow of ambient air and vacuum is applied to the inflation's air chamber during the milking process. When ambient air is allowed to enter the chamber, the vacuum inside the inflation causes the inflation to collapse around the cow's teat, squeezing the milk out of teat in a similar fashion as a baby calf's mouth massaging the teat. When the vacuum is reapplied in the chamber the flexible rubber inflation relaxes and opens up, preparing for the next squeezing cycle.

It takes the average cow three to five minutes to give her milk. Some cows are faster or slower. Slow-milking cows may take up to fifteen minutes to let down all their milk. Though milking speed is not related to the quality of milk produced by the cow, it does impact the management of the milking process. Because most milkers milk cattle in groups, the milker can only process a group of cows at the speed of the slowest-milking cow. For this reason, many farmers will group slow-milking cows so as not to stress the faster milking cows.

The extracted milk passes through a strainer and plate heat exchangers before entering the tank, where it can be stored safely for a few days at approximately 40 °F (4 °C). At pre-arranged times, a milk truck arrives and pumps the milk from the tank for transport to a dairy factory where it will be pasteurized and processed into many products. The frequency of pick up depends and the production and storage capacity of the dairy; large dairies will have milk pick-ups once per day.

Management of the Herd

Modern dairy farmers use milking machines and sophisticated plumbing systems to harvest and store the milk from the cows, which are usually milked two or three times daily. In New Zealand, some farmers seeking a better life style, are milking only once per day, trading a slight reduction in production of milk for increased leisure time. During the summer months, cows may be turned out to graze in pastures, both day and night, and are brought into the barn to be milked.

Barns may also incorporate tunnel ventilation into the architecture of the barn structure. This ventilation system is highly efficient and involves opening both ends of the structure allowing cool air to blow through the building. Farmers with this type of structure keep cows inside during the summer months to prevent heat stress, sunburn and damage to udders. During the winter months the cows may be kept in the barn, which is warmed by their collective body heat. Even in winter, the heat produced by the cattle requires the barns to be ventilated for cooling purposes. Many large, modern facilities, and particularly those in tropical areas, keep all animals inside at all times to facilitate herd management.

Farmers typically grow their own food for their cattle. Crops grown may include corn, alfalfa, timothy, wheat, oats, sorghum and clover. These plants are often processed after harvest to preserve or improve nutrient value and prevent spoiling. Corn, alfalfa, wheat, oats, and sorghum crops are often anaerobically fermented to create silage. Many crops such as alfalfa, timothy, oats, and clover are allowed to dry in the field after cutting before being baled into hay.

In the southern hemisphere such as in Australia and New Zealand, cows spend most of their lives outside on pasture, although they may receive supplementation during periods of low pasture availability. Typical supplementary feeds in Australasia are hay, silage or ground maize. The trend in New Zealand is towards feeding cows on a concrete pad to prevent loss of feed by trampling. In New Zealand slower growing winter pasture is rationed. It is carefully controlled by light weight portable electric break feeding fences run on mains power that can be easily repositioned.

Concerns

Animal Waste from Large Cattle Dairies

Dairy CAFO—EPA

As measured in phosphorus, the waste output of 5,000 cows roughly equals a municipality of 70,000 people. In the U.S., dairy operations with more than 1,000 cows meet the EPA definition of a CAFO (Concentrated Animal Feeding Operation), and are subject to EPA regulations. For example, in the San Joaquin Valley of California a number of dairies have been established on a very large scale. Each dairy consists of several modern milking parlor set-ups operated as a single enterprise. Each milking parlor is surrounded by a set of 3 or 4 loafing barns housing 1,500 or 2,000 cattle. Some of the larger dairies have planned 10 or more series of loafing barns and milking parlors in this arrangement, so that the total operation may include as many as 15,000 or 20,000 cows. The milking process for these dairies is similar to a smaller dairy with a single milking parlor but repeated several times. The size and concentration of cattle creates major environmental issues associated with manure handling and disposal, which requires substantial areas of cropland (a ratio of 5 or 6 cows to the acre, or several thousand acres for dairies of this size) for manure spreading and dispersion, or several-acre methane digesters. Air pollution from methane gas associated with manure management also is a major concern. As a result, proposals to develop dairies of this size can be controversial and provoke substantial opposition from environmentalists including the Sierra Club and local activists.

The potential impact of large dairies was demonstrated when a massive manure spill occurred on a 5,000-cow dairy in Upstate New York, contaminating a 20-mile (32 km) stretch of the Black River, and killing 375,000 fish. On 10 August 2005, a manure storage lagoon collapsed releasing 3,000,000 US gallons (11,000,000 l; 2,500,000 imp gal) of manure into the Black River. Subsequently the New York Department of Environmental Conservation mandated a settlement package of $2.2 million against the dairy.

When properly managed, dairy and other livestock waste, due to its nutrient content (N, P, K), makes an excellent fertilizer promoting crop growth, increasing soil organic matter, and improving overall soil fertility and tilth characteristics. Most dairy farms in the United States are required to develop nutrient management plans for their farms, to help balance the flow of nutrients and reduce the risks of environmental pollution. These plans encourage producers to monitor all nutrients coming onto the farm as feed, forage, animals, fertilizer, etc. and all nutrients exiting the farm as product, crop, animals, manure, etc. For example, a precision approach to animal feeding results in less overfeeding of nutrients and a subsequent decrease in environmental excretion of nutrients, such as phosphorus. In recent years, nutritionists have realized that requirements for phosphorus are much lower than previously thought. These changes have allowed dairy producers to reduce the amount of phosphorus being fed to their cows with a reduction in environmental pollution.

Use of Hormones

It is possible to maintain higher milk production by supplementing cows with growth hormones known as recombinant BST or rBST, but this is controversial due to its effects on animal and possibly human health. The European Union, Japan, Australia, New Zealand and Canada have banned its use due to these concerns.

In the US however, no such prohibition exists, and approximately 17.2% of dairy cows are treated in this way. The U.S. Food and Drug Administration states that no "significant difference" has been found between milk from treated and non-treated cows but based on consumer concerns several milk purchasers and resellers have elected not to purchase milk produced with rBST.

Animal Welfare

The practice of dairy production in a factory farm environment has been criticized by animal welfare activists. Some of the ethical complaints regarding dairy production cited include how often the dairy cattle must remain pregnant, the separation of calves from their mothers, how dairy cattle are housed and environmental concerns regarding dairy production.

The production of milk requires that the cow be in lactation, which is a result of the cow having given birth to a calf. The cycle of insemination, pregnancy, parturition, and lactation, followed by a "dry" period of about two months of forty-five to fifty days, before calving which allows udder tissue to regenerate. A dry period that falls outside this time frame can result in decreased milk production in subsequent lactation.

An important part of the dairy industry is the removal of the calves off the mother's milk after the three days of needed colostrum, allowing for the collection of the milk produced. On some dairies, in order for this to take place, the calves are fed milk replacer, a substitute for the whole milk produced by the cow. Milk replacer is generally a powder, which comes in large bags, and is added to precise amounts of water, and then fed to the calf via bucket or bottle. However, not all dairies use milk replacer - some continue to feed calves milk from the cows in the milking herd. Some dairies even pasteurize extra milk from the main herd to feed calves.

Milk replacers are classified by three categories: protein source, protein/fat (energy) levels, and medication or additives (e.g. vitamins and minerals). Proteins for the milk replacer come from different sources; the more favorable and more expensive all milk protein (e.g. whey protein- a by-product of the cheese industry) and alternative proteins including soy, animal plasma and wheat gluten. The ideal levels for fat and protein in milk replacer are 10-28% and 18-30%, respectively. The higher the energy levels (fat and protein), the less starter feed (feed which is given to young animals) the animal will consume. Weaning can take place when a calf is consuming at least two pounds of starter feed a day and has been on starter for at least three weeks. Milk replacer has climbed in cost US$15–20 a bag in recent years, so early weaning is economically crucial to effective calf management.

Because of the danger of infection to humans, it is important to maintain the health of milk-producing cattle. Common ailments affecting dairy cows include infectious disease (e.g. mastitis, endometritis and digital dermatitis), metabolic disease (e.g. milk fever and ketosis) and injuries caused by their environment (e.g. hoof and hock lesions).

Lameness is commonly considered one of the most significant animal welfare issues for dairy cattle, and is best defined as any abnormality that causes an animal to change its gait. It can be caused by a number of sources, including infections of the hoof tissue (e.g. fungal infections that cause dermatitis) and physical damage causing bruising or lesions (e.g. ulcers or hemorrhage of the hoof). Housing and management features common in modern dairy farms (such as concrete barn floors, limited access to pasture and suboptimal bed-stall design) have been identified as contributing risk factors to infections and injuries. New dairy farms being built now include non-slip flooring and other features designed to minimize risk to cows when moving between pens and to the milking parlor.

Market

Worldwide

There is a great deal of variation in the pattern of dairy production worldwide. Many countries which are large producers consume most of this internally, while others (in particular New Zealand), export a large percentage of their production. Internal consumption is often in the form of liquid milk, while the bulk of international trade is in processed dairy products such as milk powder.

Holstein cows on a dairy farm, Comboyne, New South Wales

The milking of cows was traditionally a labor-intensive operation and still is in less developed countries. Small farms need several people to milk and care for only a few dozen cows, though for many farms these employees have traditionally been the children of the farm family, giving rise to the term "family farm".

Dairy farm in Võru Parish, Estonia

Advances in technology have mostly led to the radical redefinition of "family farms" in industrialized countries such as Australia, New Zealand, and the United States. With farms of hundreds of cows producing large volumes of milk, the larger and more efficient dairy farms are more able to weather severe changes in milk price and operate profitably, while "traditional" very small farms generally do not have the equity or cash flow to do so. The common public perception of large corporate farms supplanting smaller ones is generally a misconception, as many small family farms expand to take advantage of economies of scale, and incorporate the business to limit the legal liabilities of the owners and simplify such things as tax management.

Before large scale mechanization arrived in the 1950s, keeping a dozen milk cows for the sale of milk was profitable. Now most dairies must have more than one hundred cows being milked at a time in order to be profitable, with other cows and heifers waiting to be "freshened" to join the milking herd . In New Zealand the average herd size, for the 2009/2010 season, is 376 cows.

Worldwide, the largest milk producer is India (more than 55% buffalo milk), the largest cow milk exporter is New Zealand, and the largest importer is China. The European Union with its present 28 member countries produced 158,800,000 metric tons (156,300,000 long tons; 175,000,000 short tons) in 2013(96.8% cow milk), the most by any politico-economic union.

Rank	Country	Production (10^6 kg/y)
\multicolumn{3}{c}{World total milk production in 2009 FAO statistics (including cow/buffalo/goat/sheep/camel milk)}		
	World	696,554
1	India	110,040
2	United States	85,859
3	China	40,553
4	Pakistan	34,362
5	Russia	32,562
6	Germany	28,691
7	Brazil	27,716
8	France	24,218
9	New Zealand	15,217
10	United Kingdom	13,237
11	Italy	12,836
12	Turkey	12,542
13	Poland	12,467
14	Ukraine	11,610
15	Netherlands	11,469
16	Mexico	10,931
17	Argentina	10,500
18	Australia	9,388
19	Canada	8,213
20	Japan	7,909

European Union

Production building at a dairy farm in Norway.

The European Union with its present 27 member countries is the largest milk producer in the world. The largest producers within the EU are Germany and France.

Dairy production in the EU is heavily distorted due to the Common Agricultural Policy – being subsidized in some areas, and subject to production quotas in other.

European total milk production in 2009 FAO statistics (including cow/goat/sheep/buffalo milk)		
Rank	Country	Production (10^6 kg/y)
	European Union (all 27 countries)	153,033
1	Germany	28,691
2	France	24,218
3	United Kingdom	13,237
4	Italy	12,836
5	Poland	12,467
6	Netherlands	11,469
7	Spain	7,252
8	Romania	5,809
9	Ireland	5.373
10	Denmark	4,814

Israel

The dairy farm on Sa'ad was the Israeli leader in 2011 for productivity with an average of 13,785 litres (3,032 imp gal; 3,642 US gal) per head that year. A dairy cow named Kharta, was the world record holder giving 18,208 litres (4,005 imp gal; 4,810 US gal) liters of milk. The 954 Israeli dairy farms achieved a world leading average production of 11,775 litres (2,590 imp gal; 3,111 US gal) a year per head, while the national average per head was 10,336 litres (2,274 imp gal; 2,730 US gal). Israeli consumption is lower than other western countries with an average of 180 litres (40 imp gal; 48 US gal) per person.

United States

In the United States, the top five dairy states are, in order by total milk production; California, Wisconsin, New York, Idaho, and Pennsylvania. Dairy farming is also an important industry in Florida, Minnesota, Ohio and Vermont. There are 65,000 dairy farms in the United States.

Pennsylvania has 8,500 farms with 555,000 dairy cows. Milk produced in Pennsylvania yields an annual revenue of about US$1.5 billion.

Milk prices collapsed in 2009. Senator Bernie Sanders accused Dean Foods of controlling 40% of the country's milk market. He has requested the United States Department of Justice to pursue an anti-trust investigation. Dean Foods says it buys 15% of the country's raw milk. In 2011, a federal judge approved a settlement of $30 million to 9,000 farmers in the Northeast.

Herd size in the US varies between 1,200 on the West Coast and Southwest, where large farms are commonplace, to roughly 50 in the Midwest and Northeast, where land-base is a significant limiting factor to herd size. The average herd size in the U.S. is about one hundred cows per farm but the median size is 900 cows with 49% of all cows residing on farms of 1000 or more cows.

Dairy Cattle

A Holstein cow with prominent udder and less muscle than is typical of beef breeds

Dairy cattle (also called dairy cows or milk cows) are cattle cows bred for the ability to produce large quantities of milk, from which dairy products are made. Dairy cows generally are of the species *Bos taurus*.

Historically, there was little distinction between dairy cattle and beef cattle, with the same stock often being used for both meat and milk production. Today, the bovine industry is more specialized and most dairy cattle have been bred to produce large volumes of milk. The United States dairy herd produced 83.9 billion kg (185 billion lbs) of milk in 2007, up from 52.6 billion kg (116 billion lbs) in 1950, yet there were only about 9 million cows on U.S. dairy farms—about 13 million fewer than there were in 1950.

Management

Cows on a dairy farm in Maryland, U.S.

Dairy cows may be found either in herds or dairy farms where dairy farmers own, manage, care for, and collect milk from them, or on commercial farms. Herd sizes vary around the world depending on landholding culture and social structure. The United States has 9 million cows in 75,000 dairy herds, with an average herd size of 120 cows. The number of small herds is falling rapidly with the 3,100 herds with over 500 cows producing 51% of U.S. milk in 2007. The United Kingdom dairy herd overall has nearly 1.5 million cows, with about 100 head reported on an average farm. In New Zealand, the average herd has more than 375 cows, while in Australia, there are approximately 220 cows in the average herd.

To maintain lactation, a dairy cow must be bred and produce calves. Depending on market conditions, the cow may be bred with a "dairy bull" or a "beef bull." Female calves (heifers) with dairy breeding may be kept as replacement cows for the dairy herd. If a replacement cow turns out to be a substandard producer of milk, she then goes to market and can be slaughtered for beef. Male calves can either be used later as a breeding bull or sold and used for veal or beef. Dairy farmers usually begin breeding or artificially inseminating heifers around 13 months of age. A cow's gestation period is approximately nine months. Newborn calves are removed from their mothers quickly, usually within three days, as the mother/calf bond intensifies over time and delayed separation can cause extreme stress on both cow and calf.

Domestic cows can live to 20 years; however, those raised for dairy rarely live that long, as the average cow is removed from the dairy herd around age four and marketed for beef. In 2014, approximately 9.5% of the cattle slaughtered in the U.S. were culled dairy cows: cows that can no longer be seen as an economic asset to the dairy farm. These animals may be sold due to reproductive problems or common diseases of milk cows such as mastitis and lameness.

Calf

Market calves are generally sold at two weeks of age and bull calves may fetch a premium over heifers due to their size, either current or potential. Calves may be sold for veal, or for one of several types of beef production, depending on available local crops and markets. Such bull calves may be castrated if turnout onto pastures is envisaged, in order to render the animals less aggressive. Purebred bulls from elite cows may be put into progeny testing schemes to find out whether they might become superior sires for breeding. Such animals may become extremely valuable.

Most dairy farms separate calves from their mothers within a day of birth to reduce transmission of disease and simplify management of milking cows. Studies have been done allowing calves to remain with their mothers for 1, 4, 7 or 14 days after birth. Cows whose calves were removed longer than one day after birth showed increased searching, sniffing and vocalizations. However, calves allowed to remain with their mothers for longer periods showed weight gains at three times the rate of early removals as well as more searching behavior and better social relationships with other calves.

After separation, some young dairy calves subsist on commercial milk replacer, a feed based on dried milk powder. Milk replacer is an economical alternative to feeding whole milk because it is cheaper, can be bought at varying fat and protein percentages, and is typically less contaminated than whole milk when handled properly. Some farms pasteurize and feed calves milk from the cows in the herd instead of using replacer. A day-old calf consumes around 5 liters of milk per day.

Bull

A bull calf with high genetic potential may be reared for breeding purposes. It may be kept by a dairy farm as a herd bull, to provide natural breeding for the herd of cows. A bull may service up to 50 or 60 cows during a breeding season. Any more and the sperm count will decline, leading to cows "returning to service" (to be bred again). A herd bull may only stay for one season since over two years old their temperament becomes too unpredictable.

Bull calves intended for breeding commonly are bred on specialized dairy breeding farms, not production farms. These farms are the major source of stocks for artificial insemination.

Milk Production Levels

Dairy Cows, Collins Center, New York, 1999

A cow will produce large amounts of milk over its lifetime. Certain breeds produce more milk than others; however, different breeds produce within a range of around 6,800 to 17,000 kg (15,000 to

37,500 lbs) of milk per lactation. The Holstein Friesian is the main breed of dairy cattle in Australia, and said to have the "world's highest" productivity, at 10000L of milk per year. The average for a single dairy cow in the US in 2007 was 9164.4 kg (20,204 lbs) per year, excluding milk consumed by her calves, whereas the same average value for a single cow in Israel was reported in the Philippine press to be 12,240 kg in 2009. Production levels peak at around 40 to 60 days after calving. The cow is then bred. Production declines steadily afterwards, until, at about 305 days after calving, the cow is 'dried off', and milking ceases. About sixty days later, one year after the birth of her previous calf, a cow will calve again. High production cows are more difficult to breed at a one-year interval. Many farms take the view that 13 or even 14 month cycles are more appropriate for this type of cow.

Dairy cows may continue to be economically productive for many lactations. In most cases, 10 lactations are possible. The chances of problems arising which may lead to a cow being culled are high, however; the average herd life of US Holstein is today fewer than 3 lactations. This requires more herd replacements to be reared or purchased. Over 90% of all cows are slaughtered for 4 main reasons:

- Infertility - failure to conceive and reduced milk production.

  Cows are at their most fertile between 60 and 80 days after calving. Cows remaining "open" (not with calf) after this period become increasingly difficult to breed, which may be due to poor health. Failure to expel the afterbirth from a previous pregnancy, luteal cysts, or metritis, an infection of the uterus, are common causes of infertility.

- Mastitis - a persistent and potentially fatal mammary gland infection, leading to high somatic cell counts and loss of production.

  Mastitis is recognized by a reddening and swelling of the infected quarter of the udder and the presence of whitish clots or pus in the milk. Treatment is possible with long-acting antibiotics but milk from such cows is not marketable until drug residues have left the cow's system, also called withdrawal period.

- Lameness - persistent foot infection or leg problems causing infertility and loss of production.

  High feed levels of highly digestible carbohydrate cause acidic conditions in the cow's rumen. This leads to Laminitis and subsequent lameness, leaving the cow vulnerable to other foot infections and problems which may be exacerbated by standing in faeces or water soaked areas.

- Production - some animals fail to produce economic levels of milk to justify their feed costs.

  Production below 12 to 15 litres of milk per day is not economically viable.

Cow longevity is strongly correlated with production levels. Lower production cows live longer than high production cows, but may be less profitable. Cows no longer wanted for milk production are sent to slaughter. Their meat is of relatively low value and is generally used for processed meat. Another factor affecting milk production is the stress the cow is faced with. Psychologists at the University of Leicester, UK, analyzed the musical preference of milk cows and found out that music actually influences the dairy cow's lactation. Calming music can improve milk yield, probably because it reduces stress and relaxes the cows in much the same way as it relaxes humans.

By-products

By-products of milk include butterfat, cream, curds, and whey. Butterfat is the fat in milk. The cream is the yellowish part of the milk. The cream contains 18–40% butterfat. Whey is the watery part of the milk.

Reproduction

Since the 1950s, artificial insemination (AI) is used at most dairy farms; these farms may keep no bull. Advantages of using AI include its low cost and ease compared to maintaining a bull, ability to select from a large number of bulls to match the anticipated market for the resulting calves, and predictable results.

More recently, embryo transfer has been used to enable the multiplication of progeny from elite cows. Such cows are given hormone treatments to produce multiple embryos. These are then 'flushed' from the cow's uterus. 7-12 embryos are consequently removed from these donor cows and transferred into other cows who serve as surrogate mothers. The result will be between 3 and 6 calves instead of the normal single, or rarely, twins.

Hormone Use

Hormone treatments are sometimes given to dairy cows in some countries to increase reproduction and to increase milk production.

The hormones are used to produce multiple embryos have to be administered at specific times to dairy cattle to induce ovulation. Frequently, for economic considerations, these drugs are also used to synchronise a group of cows to ovulate simultaneously. The hormones prostaglandin, gonadotropin-releasing hormone, and progesterone are used for this purpose and sold under the brand names Lutalyse, Cystorelin, Estrumate, Estroplan, Factrel, Prostamate, Fertagyl, Insynch, and Ovacyst. They may be administered by injection.

About 17% of dairy cows in the United States are injected with Bovine somatotropin, also called recombinant bovine somatotropin (rBST), recombinant bovine growth hormone (rBGH), or artificial growth hormone. The use of this hormone increases milk production from 11%–25%. The U.S. Food and Drug Administration (FDA) has ruled that rBST is harmless to people. The use of rBST is banned in Canada, parts of the European Union, as well as Australia and New Zealand.

Nutrition

Nutrition plays an important role in keeping cattle healthy and strong. Implementing an adequate nutrition program can also improve milk production and reproductive performance. Nutrient requirements may not be the same depending on the animal's age and stage of production.

Dairy cattle at feeding time

Forages, which refer especially to hay or straw, are the most common type of feed used. Cereal grains, as the main contributors of starch to diets, are important in meeting the energy needs of dairy cattle. Barley is one example of grain that is extensively used around the world. Barley is grown in temperate to subarctic climates, and it is transported to those areas lacking the necessary amounts of grain. Although variations may occur, in general, barley is an excellent source of balanced amounts of protein, energy, and fiber.

Ensuring adequate body fat reserves is essential for cattle to produce milk and also to keep reproductive efficiency. However, if cattle get excessively fat or too thin, they run the risk of developing metabolic problems and may have problems with calving. Scientists have found that a variety of fat supplements can benefit conception rates of lactating dairy cows. Some of these different fats include oleic acids, found in canola oil, animal tallow, and yellow grease; palmitic acid found in granular fats and dry fats; and linolenic acids which are found in cottonseed, safflower, sunflower, and soybean. It is also important to note that proper levels of fat also improve cattle longevity.

Using by-products is one way of reducing the normally high feed costs. However, lack of knowledge of their nutritional and economic value limits their use. Although the reduction of costs may be significant, they have to be used carefully because animal may have negative reactions to radical changes in feeds, (e.g. fog fever). Such a change must then be made slowly and with the proper follow up.

Pesticide Use

A survey of the primary dairy producing areas in the US indicated that 13 percent of lactating animals were treated with insecticides permethrin, pyrethrin, coumaphos, and dichlorvos primarily by daily or every-other-day coat sprays. Workers, particularly in stanchion barns, may be exposed to higher than recommended amounts of these pesticides.

Breeds

According to the Purebred Dairy Cattle Association, PDCA, there are 7 major dairy breeds in the United States. These are: Holstein, Brown Swiss, Guernsey, Ayrshire, Jersey, Red and White, and Milking Shorthorn.

Holstein cows have distinct white and black markings. Holstein cows are the biggest of all U.S. dairy breeds. A full mature Holstein cow usually weighs around 1,500 pounds and is 58 inches tall at the shoulder. They are known for their outstanding milk production among the main breeds of dairy cattle. An average Holstein cow produces around 23,000 pounds of milk each lactation. Of the 9 million dairy cows in the U.S., approximately 90% of them are of the Holstein descent.

Brown Swiss cows are widely accepted as the oldest dairy cattle breed, originally coming from a part of northeastern Switzerland. Some experts think that the modern Brown Swiss skeleton is similar to one found that looks to be from around the year 4000 B.C. Also, there is evidence that monks started breeding these cows about 1000 years ago.

The Ayrshire breed first originated in the County of Ayr in Scotland. It became regarded as a well established breed in 1812. The different breeds that were crossed to form the Ayrshire are not exactly known. However, there is evidence that several breeds were crossed with the native cattle to create the breed.

Guernsey cows originated just off the coast of France on the small Isle of Guernsey. The breed was first known as a separate breed around 1700. Guernseys are known for their ability to produce very high quality milk from grass. Also, the term "Golden Guernsey" is very common as Guernsey cattle produce rich, yellow milk rather than the standard white milk other cow breeds produce.

Unlike the other cow breeds, it is not quite known where or how Jersey cows originated. However, most signs point to the fact that they came from an area of France close to Normandy. These cows more than likely came from this area in France to the Jersey area of the United States, hence the name "Jersey" cows. It also is cited that the Jersey cows probably had a major role in the success of the Jersey area. Jersey cows, according to available data, have been in the UK area since about the year 1741. When they were in this area, they were not known as Jerseys, but rather as Alderneys. The period between 1860 and around 1914 was the best time for Jerseys. In this time span, many countries other than the United States started importing this breed, including Canada, South Africa, and New Zealand, among others. Among the smallest of the dairy breeds, the average Jersey cow matures at approximately 900 pounds, with a typical weight range between 800 and 1,200 pounds. The milk of the Jersey boasts the highest milk fat content. This high fat content means the milk is often used for making ice cream and cheeses. According to the American Jersey Cattle Association, Jerseys are found on 20 percent of all US dairy farms and are the primary breed on about 4 percent of dairies. According to North Dakota State University, the fat content of the Jersey cow is nearly 5 percent—4.9 percent, to be exact. It's also the highest in protein, at 3.8 percent.

Amongst the Bos indicus, the most popular dairy breed in the world is Sahiwal of the Indian subcontinent. It does not give as much milk as the Taurine breeds, but it is by far the most suitable breed for warmer climates. Australian Friesian Sahiwal and Australian Milking Zebu have been developed in Australia using Sahiwal genetics. Gir, another of the Bos Indicus breeds, has been improved in Brazil for its milk production and is widely used there for dairy.

Animal Welfare

Animal welfare refers to both the physical and mental state of an animal, and how it is coping with its situation. An animal is considered in a good state of welfare if it is able to express its innate behaviour, comfortable, healthy, safe, well nourished, and is not suffering from negative states such as distress, fear and pain. Good animal welfare requires disease prevention and veterinary treatment, appropriate shelter, management, nutrition, humane handling, transport and eventually, humane slaughter.

Proper animal handling, or stockmanship, is crucial to dairy animals' welfare as well as the safety of their handlers. Improper handling techniques can stress cattle leading to impaired production and health, such as increased slipping injuries. Additionally, the majority of nonfatal worker injuries on a dairy farm are from interactions with cattle. Dairy animals are handled on a daily basis for a wide variety of purposes including health-related management practices and movement from freestalls to the milking parlor. Due to the prevalence of human-animal interactions on dairy farms, researchers, veterinarians, and farmers alike have focused on furthering our understanding of stockmanship and educating agriculture workers. Stockmanship is a complex concept that involves the timing, positioning, speed, direction of movement, and sounds and touch of the handler. A recent survey of Minnesota dairy farms revealed that 42.6% of workers learned stockmanship techniques from a family members, and 29.9% had participated in stockmanship training. However as the growing U.S. dairy industry increasingly relies on an immigrant workforce, stockmanship training and education resources will become more pertinent. Clearly communicating and managing a large culturally diverse workforce brings new challenges such as language barriers and time limitations. Organizations like the Upper Midwest Agriculture Safety and Health Center (UMASH) offer resources such as bilingual training videos, fact sheets, and informational posters for dairy worker training. Additionally the Beef Quality Assurance Program offer seminars, live demonstrations, and online resources for stockmanship training.

The practice of dairy production in a factory farm environment has been criticized by animal rights activists. Some of the ethical reasons regarding dairy production cited include how often the dairy cattle are impregnated, the separation of calves from their mothers, and the fact that the cows are considered "spent" and culled at a relatively young age, as well as environmental concerns regarding dairy production.

The production of milk requires that the cow be in lactation, which is a result of the cow having given birth to a calf. The cycle of insemination, pregnancy, parturition, and lactation is followed by a "dry" period of about two months before calving, which allows udder tissue to regenerate. A dry period that falls outside this time frames can result in decreased milk production in subsequent lactation. Dairy operations therefore include both the production of milk and the production of calves. Bull calves are either castrated and raised as steers for beef production or veal.

Animal rights groups such as Mercy for Animals also raise welfare concerns by citing undercover footage showing abusive practices at factory farms.

Milking Pipeline

A milking pipeline or milk pipeline is a component of a dairy farm animal-milking operation which is used to transfer milk from the animals to a cooling and storage bulk tank.

Setup

In small dairy farms with less than 100 cows, goats or sheep, the pipeline is installed above the animals' stalls and they are then are milked in sequence by moving down the row of stalls. The milking machine is a lightweight transportable hose assembly which is plugged into sealed access ports along the pipeline.

In the United States, for farmers who participate in the voluntary Dairy Herd Improvement Association, approximately once a month the milk volume from each animal is measured using additional portable metering devices inserted between the milker and the pipeline.

In large dairy farms with more than 100 animals, the pipeline is installed within a milking parlor that the animals walk through in order to be milked at fixed stations. Because the machine is stationary, it can include additional fixed equipment such as computerized milk-metering systems to measure volume, which would be cumbersome to use with portable milkers.

In both cases the pipeline is constructed out of stainless steel, which does not easily corrode and is resistant to most chemicals, though larger operations may use larger-diameter pipes in order to handle greater milk volumes.

Transfer from Pipeline to Bulk Tank

There is usually a transition point to move the milk from the pipeline under vacuum to the bulk tank, which is at normal atmospheric pressure. This is done by having the milk flow into a receiver bowl or globe, which is a large hollow glass container with electronic liquid-detecting probes in the center. As the milk rises to a certain height in the bowl, a transfer pump is used to push it through a one-way check valve and into a pipe that transfers it to the bulk tank. When the level has dropped far enough in the bowl, the transfer pump turns off. Without the check valve, the milk in the bulk tank could be sucked back into the receiver bowl when the pump is not running.

In the event of electronics or pump failure, there is also usually a secondary bowl attached to the top of receiver bowl, which contains a float and a diaphragm valve. If the main receiver bowl overflows due to pump failure, the rising milk lifts the float in the secondary bowl, which will cut off vacuum to the entire milk pipeline and will prevent the milk or wash water from being sucked into the vacuum pump.

Some milk handling systems eliminate the receiver bowl and transfer pump by having rubber seals on the bulk tank covers, to permit the entire tank to be under vacuum until milking is finished. Milk can then just flow directly by gravity from the pipeline into the bulk tank.

Pipeline Cleaning

320x240, 170 kilobit video of the pipeline cleaning process for a small 35-cow dairy farm, that has a traditional stanchion barn with haymow. The automatic washing system shown is a 1970s Bender Machine Works "Trol-O-Matic 5570", and the pipeline receiver and pump were made by Sta-Rite.

The pipeline and all milk handling systems are cleaned after every milking session using a washing system that first rinses out the remaining milk and then flushes cleaning solution through the piping to kill bacteria and remove *milkstone*, a layer of scale mainly formed by cations like calcium and magnesium. The entire washing mechanism is operated very much like a household dishwasher with an automatic fill system, soap dispenser, and automatic drain opener.

The pipeline is usually set up so that the vacuum in the system that lifts milk up can also be used to drive the cleaning process. Rather than having a single line run to the bulk tank, typically a pair of lines transport milk by gravity flow to the receiver bowl and transfer pump. The high ends of these two lines are joined together to form a complete loop back to the receiver bowl.

Cleaning is accomplished by inserting a choke plug into one of the lines leading to the transfer pump, and sucking large volumes of water from a wash-water supply tank into the choked line. This choke plug is mounted on a rod, and is inserted into the line before cleaning, and pulled out for regular milking. Due to the choke, the water, which is sufficient to completely fill the pipe, is sucked up one side of the pipeline, over the high point joining the two pipeline sections, and then flows back to the receiver bowl and transfer pump through the unchoked line. The transfer pump is then used to move the cleaning solution from the receiver bowl back to the wash-water supply tank to restart the process.

Typically, the inlet ports on the receiver globe are designed so that large slugs of wash water moving at high speed will enter on a tangent to the sides of the globe and rapidly spin around inside to assist in vigorous cleaning of the globe's interior. It is normal for wash water to overflow out the top of the globe and for some wash water to be sucked into the overflow chamber to also flush it out. During cleaning the bottom of the overflow chamber is connected to a drain channel on the receiver globe to permit water to flow out.

For the small-farm pipeline, portable milkers are inserted into this cleaning loop usually by sucking the cleaning solution out of the wash supply tank through the milker claw and outputting from the milker hoses into the choked end of the line. When the water returns to the receiver bowl, the transfer pump returns the water back to the milker's water pickup tank.

Automatic Milking

Automatic milking is the milking of dairy animals, especially of dairy cattle, without human labour. Automatic milking systems (AMS), also called voluntary milking systems (VMS), were developed in the late 20th century. They have been commercially available since the early 1990s. The core of such systems that allows complete automation of the milking process is a type of agricultural robot. Automated milking is therefore also called robotic milking. Common systems rely on the use of computers and special herd management software.

Automation in Milking

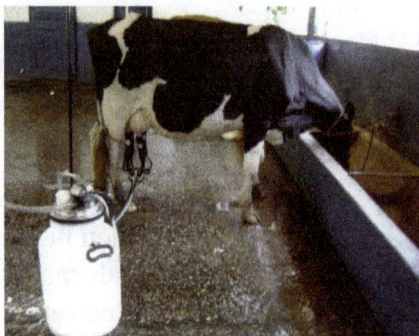

A cow and a milking machine – partial automation compared to hand milking

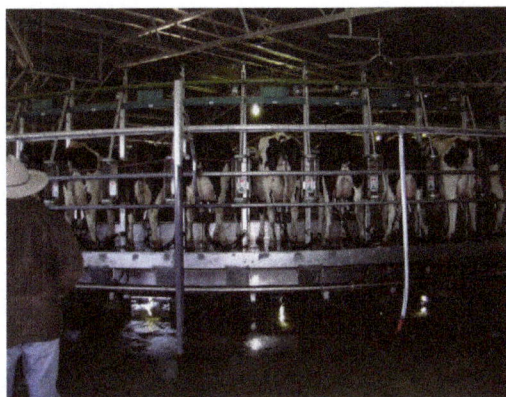

A rotary milking parlor – higher efficiency compared to stationary milking parlors, but still requiring manual labour with milking machines etc.

Basics – milking Process and Milking Schedules

The milking process is the collection of tasks specifically devoted to extracting milk from an animal (rather than the broader field of dairy animal husbandry). This process may be broken down into several sub-tasks: collecting animals before milking, routing animals into the parlour, inspection and cleaning of teats, attachment of milking equipment to teats, and often massaging the back of the udder to relieve any held back milk, extraction of milk, removal of milking equipment, routing of animals out of the parlour.

Maintaining milk yield during the lactation period (approximately 300 days) requires consistent milking intervals, usually twice daily and with maximum time spacing between milkings. In fact all activities must be scheduled around the milking process on the dairy farm. Such a milking routine imposes restrictions on time management and personal life of an individual farmer, as the farmer is committed to milking in the early morning and in the evening for seven days a week regardless of personal health, family responsibilities or social schedule. This time restriction is exacerbated for lone farmers and farm families if extra labour cannot easily or economically be obtained, and is a factor in the decline in small-scale dairy farming. Techniques such as once-a-day milking and voluntary milking have been investigated to reduce these time constraints.

Automation Progress in the 20th Century

To alleviate the labour involved in milking, much of the milking process has been automated during the 20th century: many farmers use semi-automatic or automatic cow traffic control (powered gates, etc.), the milking machine (a basic form was developed in the late 19th century) has entirely automated milk extraction, and automatic cluster removal is available to remove milking equipment after milking. Automatic teat spraying systems are available, however there is some debate over the cleaning effectiveness of these.

The final manual labour tasks remaining in the milking process were cleaning and inspection of teats and attachment of milking equipment (milking cups) to teats. Automatic cleaning and attachment of milking cups is a complex task, requiring accurate detection of teat position and a dextrous mechanical manipulator. These tasks have been automated successfully in the voluntary milking system (VMS), or automatic milking system (AMS).

Automatic Milking Systems (AMS)

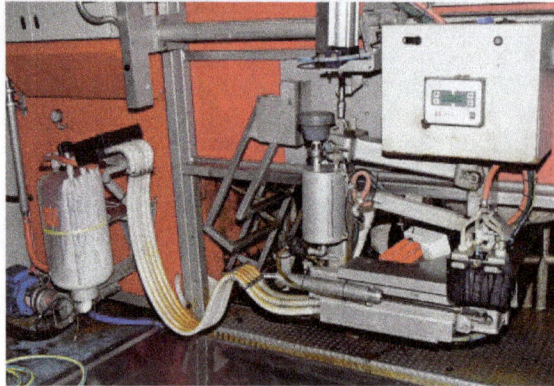

An older Lely *Astronaut* AMS unit at work (milking)

Since the 1970s, much research effort has been expended in investigating methods to alleviate time management constraints in conventional dairy farming, culminating in the development of the automated voluntary milking system. There is a video of the historical development of the milking robot at Silsoe Research Institute.

Voluntary milking allows the cow to decide her own milking time and interval, rather than being milked as part of a group at set milking times. AMS requires complete automation of the milking process as the cow may elect to be milked at any time during a 24-hour period.

The milking unit comprises a milking machine, a teat position sensor (usually a laser), a robotic arm for automatic teat-cup application and removal, and a gate system for controlling cow traffic. The cows may be permanently housed in a barn, and spend most of their time resting or feeding in the free-stall area. If cows are to be grazed as well, a selection gate is required to allow only those cows that have been milked to the outside pastures.

When the cow elects to enter the milking unit (due to highly palatable feed that she finds in the milking box), a cow ID sensor reads an identification tag (transponder) on the cow and passes the cow ID to the control system. If the cow has been milked too recently, the automatic gate system sends the cow out of the unit. If the cow may be milked, automatic teat cleaning, milking cup application, milking, and teatspraying takes place. As an incentive to attend the milking unit, concentrated feedstuffs needs to be fed to the cow in the milking unit.

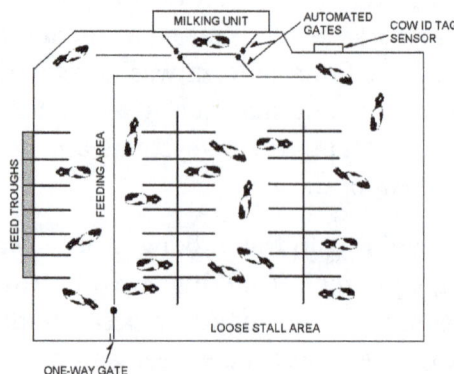

Typical VMS stall layout (*forced cow traffic* layout)

The barn may be arranged such that access to the main feeding area can only be obtained by passing the milking unit. This layout is referred to as *forced cow traffic*. Alternatively, the barn may be set up such that the cow always has access to feed, water, and a comfortable place to lie down, and is only motivated to visit the milking system by the palatable feed available there. This is referred to as *free cow traffic.*

The innovative core of the AMS system is the robotic manipulator in the milking unit. This robotic arm automates the tasks of teat cleaning and milking attachment and removes the final elements of manual labour from the milking process. Careful design of the robot arm and associated sensors and controls allows robust unsupervised performance, such that the farmer is only required to attend the cows for condition inspection and when a cow has not attended for milking.

Typical capacity for an AMS is 50–70 cows per milking unit. AMS usually achieve milking frequencies between 2 and 3 times per day, so a single milking unit handling 60 cows and milking each cow 3 times per day has a capacity of 7.5 cows per hour. This low capacity is convenient for lower-cost design of the robot arm and associated control system, as a window of several minutes is available for each cow and high-speed operation is not required.

AMS units have been available commercially since the early 1990s, and have proved relatively successful in implementing the voluntary milking method. Many of the research and developments have taken place in the Netherlands. The most farms with AMS are located in the Netherlands, and Denmark.

A new variation on the theme of robotic milking includes a similar robotic arm system, but coupled with a rotary platform, improving the number of cows that can be handled per robot arm.

Advantages

An AMS unit at work (teat cleaning)

- Elimination of labour - The farmer is freed from the milking process and associated rigid schedule, and labour is devoted to supervision of animals, feeding, etc.

- Milking consistency – The milking process is consistent for every cow and every visit, and is not influenced by different persons milking the cows. The four separate milking cups are removed individually, meaning that an empty quarter does not stay attached while the other three are finishing, resulting in less threat of injury. The newest models of automatic milkers can vary the pulsation rate and vacuum level based on milk flow from each quarter.

- Increased milking frequency – Milking frequency may increase to three times per day, however typically 2.5 times per day is achieved. This may result in less stress on the udder and increased comfort for the cow, as on average less milk is stored. Higher frequency milking increases milk yield per cow, however much of this increase is water rather than solids.

- Perceived lower stress environment – There is a perception that elective milking schedules reduce cow stress. A study found no decrease in stress between automatic and conventional milking.

- Herd management – The use of computer control allows greater scope for data collection. Such data allows the farmer to improve management through analysis of trends in the herd, for example response of milk production to changes in feedstuffs. Individual cow histories may also be examined, and alerts set to warn the farmer of unusual changes indicating illness or injury. Information gathering provides added value for AMS, however correct interpretation and use of such information is highly dependent on the skills of the user or the accuracy of computer algorithms to create attention reports.

Considerations and Disadvantages

- Higher initial cost – AMS systems cost approximately €120,000 ($190,524) per milking unit as of 2003 (presuming barn space is already available for loose-stall housing). Equipment costs decreased from $175,000 for the first stall to $158,000. Equipment costs decreased from $10,000/stall for a double-six parlor to $9000/stall for a double-ten parlor with a cost of $1200/stall for pipeline milking. Initial parlor cost was increased $5000/stall to represent a high cost parlor. Whether it is economically beneficial to invest in an AMS instead of a conventional milking parlor depends on constructions costs, investments in the milking system and costs of labour. Besides costs of labour, the availability of labour should also be taken into account. In general, an AMS is economically beneficial for smaller scale farms, and large dairies can usually operate more cheaply with a milking parlor.

- Increased electricity costs - to operate the robots, but this can be more than outweighed by reduced labour input.

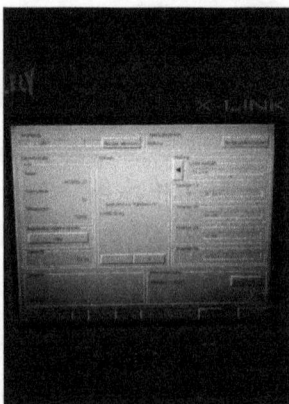

Touchscreen display of a milking robot

- Increased complexity – While complexity of equipment is a necessary part of technological advancement, the increased complexity of the AMS milking unit over conventional systems, increases the reliance on manufacturer maintenance services and possibly increasing operating costs. The farmer is exposed in the event of total system failure, relying on prompt response from the service provider. In practice AMS systems have proved robust and manufacturers provide good service networks. Because all milking cows have to visit the AMS voluntarily, the system requires a high quality of management. The system also involves a central place for the computer in the daily working routines.

- Difficult to apply in pasture systems – AMS works best in zero-grazing systems, in which the cow is housed indoors for most of the lactation period. Zero-grazing suits areas (e.g. the Netherlands) where land is at a premium, as maximum land can be devoted to feed production which is then collected by the farmer and brought to the animals in the barn. In pasture systems, cows graze in fields and are required to walk to the milking parlour. It has been found that cows tend not to attend the milking unit if the distance to walk is too great. There are currently research projects at the Dexcel facility in New Zealand, University of Sydney's FutureDairy site, and Michigan State University's Kellogg Biological Station, where cattle are on pasture and milked by AMS.

- Lower milk quality – Somatic cell count (SCC) and Plate loop count (PLC) are, respectively, measurements of the quantity of white blood cells and total number of bacteria present in a milk sample. A high SCC indicates reduced udder health (as the immune system fights some infection) and implies lower milk quality. AMS herds consistently show higher SCCs than conventionally milked herds. A high PLC indicates bacterial contamination, usually through poor sanitation or cooling and similarly implies low milk quality. High PLC in AMS may be attributed to the continuous use of milking lines (rather than twice a day in conventional systems), which reduces the time window for cleaning, and the incremental addition of milk to the bulk milk tank which may not cool efficiently at low milk levels.

- Possible increase in stress for some cows – Cows are social animals, and it has been found that due to dominance of some cows, others will be forced to milk only at night. Such behaviour is inconsistent with the perception that AM reduces stress by allowing "free choice" of milking time.

- Decreased contact between farmer and herd – Effective animal husbandry requires that the farmer be fully aware of herd condition. In conventional milking, the cows are observed before milking equipment is attached, and ill or injured cows can be earmarked for attention. Automatic milking removes the farmer from such close contact with the animal, with the possibility that illness may go unnoticed for longer periods and both milk quality and cow welfare suffer. In practice, milk quality sensors at the milking unit attempt to detect changes in milk due to infection, and farmers inspect the herd frequently. However this concern has meant that farmers are still tied to a seven-day schedule. Modern automatic milking systems attempt to rectify this problem by gathering data that would not be available in many conventional systems including milk temperature, milk conductivity, milk color including infrared scan, change in milking speed, change in milking time or milk letdown by quarter, cow's weight, cow's activity (movements), time spent ruminating, etc.

Manufacturers

- Lely (Netherlands), *Lely Astronaut AMS*

- DeLaval (Sweden), *DeLaval VMS*

- Fullwood (UK), *Merlin AMS*

- GEA Farm Technologies (Germany, formerly WestfaliaSurge), *MIone AMS*

- SAC (Denmark), purchased the Dutch manufacturer of the *Galaxy Robot AMS* in 2005, sell under the brands *SAC RDS Futureline MARK II, Insentec Galaxy Starline, BouMatic's ProFlex*

- BoumaticRobotics (NL), *MR-S1, MR-D1*

A DeLaval *VMS* unit, 2007

Bovine Somatotropin

rBST is a product primarily given to dairy cattle by injection to increase milk production.

Bovine somatotropin or bovine somatotrophin (abbreviated *bST* and *BST*), or bovine growth hormone (BGH), is a peptide hormone produced by cows' pituitary glands. Like other hormones, it is produced in small quantities and is used in regulating metabolic processes. After the biotech company Genentech discovered and patented the gene for BST in the 1970s, it became possible to synthesize the hormone using recombinant DNA technology to create recombinant bovine somatotropin (rBST), recombinant bovine growth hormone (rBGH), or artificial growth hormone.

Four large pharmaceutical companies, Monsanto, American Cyanamid, Eli Lilly, and Upjohn, developed commercial rBST products and submitted them to the US Food and Drug Administration (FDA) for approval. Monsanto was the first firm to receive approval. Other countries (Mexico, Brazil, India, Russia, and at least ten others) also approved rBST for commercial use. Monsanto licensed Genentech's patent, and marketed their product as "Posilac". In October 2008, Monsanto sold this business, in full, to Eli Lilly and Company for $300 million plus additional consideration.

rBST has not been allowed on the market in Canada, Australia, New Zealand, Japan, Israel, or the European Union since 2000. Argentina also banned the use of rBST.

The FDA, World Health Organization, and National Institutes of Health have independently stated that dairy products and meat from BST-treated cows are safe for human consumption. In the United States, public opinion led some manufacturers and retailers to market only milk that is rBST-free.

A European Union report on the animal welfare effects of rBST states that its use often results in "severe and unnecessary pain, suffering and distress" for cows, "associated with serious mastitis, foot disorders and some reproductive problems".

History

In 1937, the administration of BST was shown to increase the milk yield in lactating cows by preventing mammary cell death in dairy cattle. Until the 1980s, use of the compound was very limited in agriculture as the sole source of the hormone was from bovine carcasses. During this time, the knowledge of the structure and function of the hormone increased. With the advent of biotechnology, one of the pioneering biotech companies, Genentech, succeeded in cloning the gene for BST. Monsanto had been working along the same lines and struck a deal with Genentech in 1979 to license Genentech's patents and collaborate on development of a recombinant version of BST – a process on which Monsanto would invest $300 million. The two companies used genetic engineering to clone the BST gene into *E. coli*. The bacteria are grown in bioreactors, then broken up and separated from the rBST, which is purified to produce the injectable hormone. They published their first field trial results in 1981.

Lilly, American Cyanamid, Upjohn, and Monsanto all submitted applications to market rBST to the U.S. FDA, and the FDA completed its review of the human safety component of these applications in 1986 and found food from rBST-treated cows to be safe; however, strong public concern led to calls for more studies, investigations, and public discussions, which included an unprecedented conference on the safety of rBST in 1990 organized by the National Institutes of Health at the request of Sen. Patrick Leahy. FDA approved Monsanto's application in 1993. Monsanto launched rBST, brand-named Posilac, in 1994.

Mechanism of Action

An average dairy cow begins her lactation with a moderate daily level of milk production. This daily output increases until, at about 70 days into the lactation, production peaks. From that time until the cow is dry, production slowly decreases. This increase and decrease in production is partially caused by the count of milk-producing cells in the udder. Cell counts begin at a moderate number, increase during the first part of the lactation, then decrease as the lactation proceeds. Once lost, these cells generally do not regrow until the next lactation.

Administration of rBST or BST prior to peak production, in cows that are well fed, slows the rate at which the number of mammary cells decreases, and increases the amount of nutrients directed away from fat and toward the mammary cells, leading to an extension of peak milk production. The effects are mediated by the insulin-like growth factor (IGF) system, which is upregulated in response to BST or rBST administration in well-fed cows.

Use on Farms

From 2000-2005, the USDA National Agricultural Statistics Service survey of dairy producers found that about 17% of producers used rBST. The 2010 USDA National Agricultural Statistics Service survey of Wisconsin farms found that about 18% of dairy farms used rBST.

To apply Posilac for maximum effect, farmers are recommended to make the first application about 50 days into the cow's lactation, just before she peaks. The Posilac then sustains already-present mammary cells, limiting the rate of production decrease after production peaks. After the peak, production declines with or without application of Posilac, but declines more slowly with it than without, which permits dairy cows to produce more milk over the span of a lactation. A FAQ document created by the FDA states that, when injected into dairy cattle, the product can increase milk production by an average of more than 10% over the span of 300 days, in cows whose feed levels are increased.

Controversy

Though approved by the FDA in 1993, rBST has been immersed in controversy since the early 1980s. Part of the controversy concerns potential effects on animal health and human health.

Animal Health

Two meta-analyses have been published on rBST's effects on bovine health. Findings indicated an average increase in milk output ranging from 11%–16%, a nearly 25% increase in the risk of clinical mastitis, a 40% reduction in fertility, and 55% increased risk of developing clinical signs of lameness. The same study reported a decrease in body condition score for cows treated with rBST, though an increase in their dry matter intake occurred.

Mastitis has cost American dairy industries an estimated $1.5 to 2 billion per year in treating dairy cows.

The use of rBST increases health problems with cows, including mastitis.

In 1994, a European Union scientific commission was asked to report on the incidence of mastitis and other disorders in dairy cows and on other aspects of their welfare. The commission's statement, subsequently adopted by the European Union, stated that the use of rBST substantially increased health problems with cows, including foot problems, mastitis, and injection site reactions, impinged on the welfare of the animals, and caused reproductive disorders. The report concluded, on the basis of the health and welfare of the animals, rBST should not be used. Health Canada prohibited the sale of rBST in 1999; the external committees found, although there was no significant health risk to humans, the drug presents a threat to animal health, and, for this reason, cannot be sold in Canada.

Monsanto-sponsored trials reviewed by the FDA asked whether the use of rBST makes cows more susceptible to mastitis. According to the FDA, which used data from eight Monsanto-sponsored trials in its decision in 1993 to approve Monsanto's rBST product, the answer is yes. The data from these eight trials, which involved 487 cows, showed that during the period of rBST treatment, mastitis incidence increased by 76% in primiparous cows and by 50% for multiparous cows. Overall, the increase was 53%.

Macronutrient Composition

The overall composition of the milk including the fat, protein, and lactose content are not altered substantially by the use of rBST in dairy cows. The milk may have a slight change in fat content within the first few weeks of rBST treatment as the cow is allowed to adjust her metabolism and feed intake. The changes in the fat content have been shown to be temporary. The composition of the milk has been examined in more than 200 different experiments. Natural variation within milk is normal with or without rBST treatment in cows due to genetics, location, feed, age, and other environmental factors. Protein in milk content has also been studied and was shown to have no apparent change in rBST treated cows. The vitamins and minerals that are normally in milk were also unaltered in milk from rBST treated cows. Freezing point, pH, thermal properties, and other manufacturing characteristics of milk were shown to be the same regardless of whether it came from rBST treated cows or not.

Hormones

rBST is present in milk from both rBST-treated and untreated cows, but it is destroyed in the digestive system and even if directly injected, has not been found to have any direct effect on humans. Researchers have found that "IGF-1 in milk is not denatured by pasteurization and the extent to which intact, active IGF-1 is absorbed through the human digestive tract remains still however uncertain" implicating that an extensive study on the nature of IGF-1 in relation to rBST milk is required.

FDA rBST labeling guidelines state, "FDA is concerned that the term 'rbST free' may imply a compositional difference between milk from treated and untreated cows rather than a difference in the way the milk is produced. Without proper context, such statements could be misleading. Such unqualified statements may imply that milk from untreated cows is safer or of higher quality than milk from treated cows. Such an implication would be false and misleading".

The FDA World Health Organization, and National Institutes of Health have independently stated that dairy products and meat from rBST-treated cows are safe for human consumption. The American Cancer Society issued a report declaring, "The evidence for potential harm to humans [from rBGH milk] is inconclusive. It is not clear that drinking milk produced using rBGH significantly increases IGF-1 levels in humans or adds to the risk of developing cancer. More research is needed to help better address these concerns."

Human Health

The effects of rBGH on human health is an ongoing debate, in part due to the lack of conclusive evidence. A few of the most debated issues include:

IGF-1 is a hormone found in humans that is responsible for growth promotion, protein synthesis, and insulin actions over the lifecycle. The hormone has been shown to influence the growth of tumors in some studies and may be linked to the development of prostate, colorectal, breast, and other cancers.

IGF-1 is also found in milk (soy included). Previous research has proposed an increase of IGF-1 in rBST-treated cows, but this claim is currently not substantiated. In addition, no current evidence shows that orally consumed IGF-1 is absorbed in humans and the dietary amount is negligible when compared to what the body produces on its own. "IGF-1 in milk is not denatured (inactivated) by pasteurization. The extent to which intact, active IGF-1 is absorbed through the human digestive tract remains uncertain.

The American Cancer Society has reviewed the evidence concerning IGF-1 in milk from rBST-treated cows, and found that: "While there may be a link between IGF-1 blood levels and cancer, the exact nature of this link remains unclear. Some studies have shown that adults who drink milk have about 10% higher levels of IGF-1 in their blood than those who drink little or no milk. But this same finding has also been reported in people who drink soymilk. This suggests that the increase in IGF-1 may not be specific to cow's milk, and may be caused by protein, minerals, or some other factors in milk unrelated to rBGH. There have been no direct comparisons of IGF-1 levels in people who drink ordinary cow's milk vs. milk stimulated by rBGH. At this time, it is not clear that drinking milk, produced with or without rBGH treatment, increases blood IGF-1 levels into a range that might be of concern regarding cancer risk or other health effects.... IGF-1 concentrations are slightly higher (to variable degrees, depending on the study) in milk from cows treated with rBGH than in untreated milk. This variability is presumed to be much less than the normal range of variation of IGF-1 in cow's milk due to natural factors, but more research is needed."

Research is supportive of milk supplying vital nutrients used in childhood development. At this time, evidence does not link rBST-treated milk with adverse health outcomes for children. Several studies have looked at the relationship between type 1 diabetes and infant feeding. Environmental triggers that may elicit an autoimmune reaction is the mechanism in which is being studied. Some studies have shown early exposure to bovine milk may predispose an infant to type 1 diabetes, whereas other studies show no causality.

The American Society of Animal Science published an article in 2014 after reviewing health issues arising from the rBST debate. The article indicated "there are no new human health issues related to the use of rbST by the dairy industry. Use of rbST has no effect on the micro- and macrocomposition of milk. Also, no evidence exists that rbST use has increased human exposure to antibiotic residues in milk. Concerns that IGF-I present in milk could have biological effects on humans have been allayed by studies showing that oral consumption of IGF-I by humans has little or no biological activity. Additionally, concentrations of IGF-I in digestive tract fluids of humans far exceed any IGF-I consumed when drinking milk. Furthermore, chronic supplementation of cows with rbST does not increase concentrations of milk IGF-I outside the range typically observed for effects of farm, parity, or stage of lactation. Use of rbST has not affected expression of retroviruses in cattle or posed an increased risk to human health from retroviruses in cattle. Furthermore, risk for development of type 1 or type 2 diabetes has not increased in children or adults consuming milk and dairy products from rbST-supplemented cows. Overall, milk and dairy products provide essential nutrients and related benefits in health maintenance and the prevention of chronic diseases."

Environmental Impact

On an industry level, supplementing one million cows with rbST would result in the same amount of milk produced using 157,000 fewer cows. Farmers are, therefore, able to improve milk production with a smaller dairy population.

Some studies show that rBST-treated cows reduce the impact of greenhouse gases in comparison with conventional and organic dairy operations. Furthermore, excretion of nitrogen and phosphorus, two major environmental pollutants arising from animal agriculture, was reduced by 9.1 and 11.8%, respectively. Carbon dioxide is recognized to be the most important anthropogenic greenhouse gas, and livestock metabolism and fossil fuel consumption are the main sources of emissions from animal agriculture.

- Livestock metabolism-use of rBST in lactating cows decreases the quantity of energy and protein needed in comparison to conventional dairy operations along with reducing the total feedstuff used.

- Fossil fuel consumption-targets atmospheric pollution and resource sustainability environmental concerns. With cows treated with rBST, producing a higher milk yield reduces the feed requirement which in turn decreases with electricity for milk production and the energy required from fossil fuels for cropping. In addition, the global warming potential is reduced equivalent to removing 400,000 family cars from the road.

When conventional, conventional with rBST, and organic dairy operations are compared, 8% fewer cows are needed in an rBST-supplemented population, whereas organic production systems require a 25% increase to meet production targets. This is due to a lower milk yield per cow due to the pasture-based system which is attributed with a greater maintenance energy expenditure associated with grazing behavior.

Lawsuit Against WTVT

In 1997, the news division of WTVT (Channel 13), a Fox-owned station in Tampa, Florida, planned to air an investigative report by Steve Wilson and Jane Akre on the health risks associated with Monsanto's bovine growth hormone product, Posilac. Just before the story was to air, Fox received a letter from Monsanto saying the reporters were biased and that the story would damage the company. Fox tried to work with the reporters to address Monsanto's concerns; Akre stated that Wilson and she went through 83 rewrites over eight months. Negotiations broke down and both reporters were eventually fired. Wilson and Akre alleged the firing was for retaliation, while WTVT contended they were fired for insubordination. The reporters then sued Fox/WTVT in Florida state court under the state's whistleblower statute. In 2000, a Florida jury found that while no evidence showed Fox/WTVT had bowed to any pressure from Monsanto to alter the story, Akre, but not Wilson, was a whistleblower and was unjustly fired. She was awarded a $425,000 settlement. At the time of the decision, "the station claimed it did not bend to Monsanto's letter and wanted to air a hard-hitting story with a number of statements critical of Monsanto." Fox appealed the decision stating that under Florida law, a whistleblower can only act if "a law, rule, or regulation" has been broken and argued that the FCC's policy against distortions or misrepresentations presented as news did not fit that definition. On February 14, 2003, the appeals court overturned the verdict, finding that Akre was not a whistleblower because of the Florida "legislature's requirement that agency statements that fit the definition of a "rule" (must) be formally adopted (rules). Recognizing an uncodified agency policy developed through the adjudicative process as the equivalent of a formally adopted rule is not consistent with this policy, and it would expand the scope of conduct that could subject an employer to liability beyond what Florida's Legislature could have contemplated when it enacted the whistle-blower's statute."

Regulation

Use of the recombinant supplement has been controversial. The assessment of the United States FDA is that no significant difference between milk from treated and untreated cows. 21 other countries have also approved marketing of rBST: Brazil, Chile, Colombia, Costa Rica, Ecuador, Egypt, Guatemala, Honduras, Jamaica, Lebanon, Mexico, Panama, Pakistan, Paraguay, Peru, Salvador, South Africa, South Korea, Uruguay and Venezuela. However, regulatory bodies in several countries, such as the EU, Canada, Japan, Australia, New Zealand, and Argentina rejected Monsanto's application to sell rBST because rBST increases the risk of health problems in cows, including clinical mastitis, reduced fertility, and reduced body condition. In Canada, bulk milk products from the United States that have been produced with rBST are still allowed to be sold and used in food manufacture (cheese, yogurt, etc.).

In 1990, the European Union placed a moratorium on its sale by all member nations. It was turned into a permanent ban starting from 1 January 2000; the decision was based solely on veterinary concerns, laws, and treaties. An in-depth report published in 1999 analysed in detail the various human health risks associated with rBST.

Canada's health board, Health Canada, refused to approve rBST for use on Canadian dairies, citing concerns over animal health. The study found the occurrence of an antibody reaction, possible hypersensitivity, in a subchronic (90-day) study of rbST oral toxicity in rats that resulted in one test animal's developing an antibody response at low dose (0.1 mg/kg/day) after 14 weeks. However, the board stated, with the exception of concerns raised regarding hypersensitivity, "the panel finds no biologically plausible reason for concern about human safety if rBST were to be approved for sale in Canada."

The Codex Alimentarius Commission, a United Nations body that sets international food standards, has to date refused to approve rBST as safe. The Codex Alimentarius does not have authority to ban or approve the hormone, but its decisions are regarded as a standard and approval by the Codex would have allowed exporting countries to challenge countries with a ban on rBST before the World Trade Organization.

United States

In 1993, the product was approved for use in the U.S. by the FDA, and its use began in 1994. The product is now sold in all 50 states.

The FDA stated that food products made from rBST-treated cows are safe for human consumption, and no statistically significant difference exists between milk derived from rBST-treated and untreated cows. The FDA found BST to be biologically inactive when consumed by humans and found no biological distinction between rBST and BST. In 1990, an independent panel convened by the National Institute of Health supported the FDA opinion that milk and meat from cows supplemented with rBST is safe for human consumption.

Labeling

The FDA does not require special labels for products produced from cows given rBST, but has charged several dairies with "misbranding" its milk as having no hormones, because all milk contains hormones and cannot be produced in such a way that it would not contain any hormones. Monsanto sued Oakhurst Dairy of Maine over its use of a label which pledged not to use artificial growth hormones. The dairy stated that its disagreement was not over the scientific evidence for the safety of rBST (Monsanto's complaint about the label), but, "We're in the business of marketing milk, not Monsanto's drugs." The suit was settled when the dairy agreed to add a qualifying statement to its label: "FDA states: No significant difference in milk from cows treated with artificial growth hormones." The FDA recommends this additional labeling, but does not require it. The settlement itself caused much controversy, with anti-rBST advocates claiming that Oakhurst had capitulated in response to intimidation by a larger corporation and others claiming that Oakhurst's milk labels were in and of themselves using misleading scare tactics that deserved legal and legislative response.

Ohio

In 2008, Ohio's Department of Agriculture (ODA) banned the use of labeling in dairy products as rBST-free because it was deemed misleading to consumers. However, the International Dairy Foods Association and the Organic Trade Association claimed ODA's ban was a violation of the first amendment by not allowing consumers to decide whether they deem the milk was safe or not and filed suit against the bill. "The district court granted summary judgment in favor of Ohio, concluding that using "rBST" as a label was inherently misleading because it implies "a compositional difference between those products that are produced with rBST and those that are not."

Kansas

In 2009, the Kansas Legislature passed a bill that would have required dairies that did not use rBST to print disclaimers on their labels that stated, "The Food and Drug Administration has determined there are no significant differences between milk from cows that receive injections of the artificial hormone and milk from those that do not." The bill was vetoed in the last days of the 2009 legislative session by then-Governor Kathleen Sebelius. The legislature removed the labeling language and passed the bill without the provision.

Pennsylvania

In 2007, Pennsylvania adopted a regulation that would have banned the practice of labeling milk as derived from cows not treated with rBST. Pennsylvania's Agriculture Secretary Dennis Wolff made the following statement in support of the measure:

"Consumers are getting confused with the extra labels. They deserve a choice, and so do producers. But from the standpoint of safety, all milk is healthy milk. Our milk is a safe product. The Pennsylvania Department of Agriculture is not in a position to say use rBST or not. The key word is: choice. I used rBST from day one of its approval to the last day that I milked cows. It was an important management tool on my dairy farm. What we oppose is the negative advertising or the selling of fear. If producers are asked to give up a production efficiency, and if that efficiency nets them $3000 or $10,000 a year for their dairy farm... That's a lot of money.

This prohibition was to go into effect 1 January 2008, but after the comment period, the guidelines were adjusted to only ban "rBST-free" claims and instead allow claims that farmers had pledged not to use rBST and accompany such claims with a disclaimer such as, "No significant difference has been shown between milk derived from rbST-treated and non-rbST-treated cows."

Response from Milk Producers and Retailers

In response to concerns from consumers and advocacy groups about milk from cows treated with rBST, some dairies, retailers, and restaurants have published policies on use of rBST in production of milk products they sell, while others offer some products or product lines that are labelled "rBST-free" or the like. Other dairies and industry groups have worked to assure the public that milk from rBST-treated cows is safe.

- Costco has no overall rBST policy, but sells brands such as Kirkland with labels pledging that no rBST was used in milk production.

- Wal-Mart announced in March 2008 that its private-label Great Value milk will be "sourced exclusively from cows that have not been treated with artificial growth hormones like recombinant bovine somatotropin (rBST)"

- Kroger announced in April 2007, "it will complete the transition of milk it processes and sells in its stores to a certified rBST-free supply by February 2008."

- Dean Foods has no overall rBST policy, but has brands, such as Oak Farms, with labels pledging that no rBST was used in milk production.

- Winder Farms, a home delivery dairy and grocer in Utah and Nevada, sells milk from rBST-free cows.

- Guernsey Farms, a dairy farm and distributor located in Northville, Michigan, sells and distributes rBST-free dairy products in southeastern Michigan. Its milk has been labeled rBST-free for a number of years.

- Safeway in the northwestern United States stopped buying from dairy farmers who use rBST in January 2007. The two Safeway plants produce milk for all Safeway stores in Oregon, southwest Washington, and parts of northern California. Safeway's plant in San Leandro, California, has been rBST-free since 2005.

- Chipotle Mexican Grill announced in June 2012 that it will serve rBST-free sour cream at its restaurants.

- Publix supermarket chain states on its website: "Publix milk is rBST-free. (No added artificial hormones.) However, the FDA has stated that no significant difference has been shown between milk derived from rbST-treated and non-rbST-treated cows."

- Braum's, a dairy and ice cream retailer in the midwest with a private herd, says on its website that it does not administer rBST to its cows.

- Starbucks's website, as of August 2012, has no statement about use of milk from cows treated with rBST. For example, its animal welfare policy is silent on the issue. It announced in January 2008 that it would no longer sell milk from cows treated with rBST in its stores in the U.S. The Organic Consumers Association, an advocacy group, claimed that Starbucks' change was due to their advocacy work.

- Ben & Jerry's ice cream uses milk and cream from dairy farms that have pledged not to use rBST.

- Oakhurst Dairy stopped using rBST in 2003 but still uses natural BST

- Tillamook County Creamery Association, a co-operative made up of 110 dairy farms, indicates on its website that its cows are not treated with hormones.

- Yoplait, in 2009, General Mills announced it would stop using milk from cows treated with rBST, and stated, "While the safety of milk from cows treated with rBST is not at issue, our consumers were expressing a preference for milk from cows not treated with rBST, and we responded."

- Upstate Niagara Cooperative's entire milk supply is rBST-free.

- Byrne Dairy requests that its farmers pledge not to use rBST; its entire fluid milk, cream, ice cream, and butter manufacture is sourced exclusively from such pledging farmers.

- Trader Joe's states, "Trader Joe's brand products contain NO ... dairy ingredients from rBST sources"

In reaction to these trends, in early 2008, a pro-rBST advocacy group called American Farmers for the Advancement and Conservation of Technology (AFACT), made up of dairies and originally affiliated with Monsanto, formed and began lobbying to ban such labels. AFACT stated that "absence" labels can be misleading and imply that milk from cows treated with rBST is inferior. The organization was dissolved in 2011.

The International Dairy Foods Association has compiled a list, last updated in 2009, of state regulations in the U.S. for referencing use of growth hormones on milk labels.

Response from Health Organizations

- The American Cancer Society "has no formal position regarding rBGH".

- The American Public Health Association (APHA) Policy #2000-11, "The Precautionary Principle and Children's Health, "encourages precautionary action to prevent potential harm to fetuses, infants, and children [from the continued manufacture and use of substances], even if some cause-and-effect relationships have not been established with scientific certainty."

- The American Nurses Association supports the development of national and state laws, regulations, and policies that specifically reduce the use of rBGH or rBST in milk and dairy production in the United States.

Response from Political Organizations

- Oregon Physicians for Social Responsibility recommends buying products from cows not injected with recombinant bovine growth hormone (rBGH or rBST).

- Health Care Without Harm opposes the use of recombinant bovine growth hormone due to its adverse impacts on animals and potential harm to humans.

Dairy Product

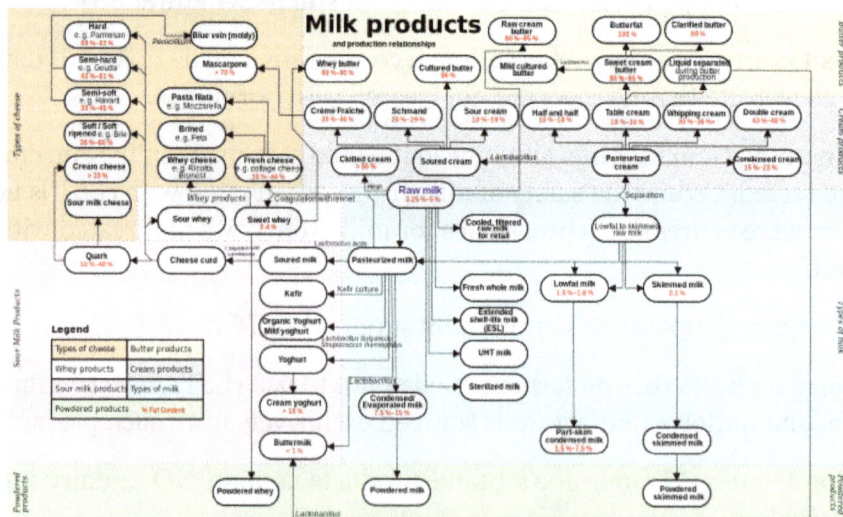

Milk products and production relationships

Dairy products or milk products are food produced from the milk of mammals, primarily cattle, water buffaloes, goats, sheep, and camels. A facility that processes milk into items like yogurt or cheese is a dairy or dairy factory. Dairy products are widely consumed worldwide, except for most of East and Southeast Asia and parts of central Africa.

Types of Dairy Products

A selection of three common dairy products made by a South African dairy company: a box of full cream, long life milk, a bottle of strawberry drinking yogurt, and a carton of passion fruit yogurt

The milk products of the Water buffaloes (super carabaos, Philippine Carabao Center)

- Milk after optional homogenization, pasteurization, in several grades after standardization of the fat level, and possible addition of the bacteria *Streptococcus lactis* and *Leuconostoc citrovorum*

    o *Crème fraîche*, slightly fermented cream

        □ Clotted cream, thick, spoonable cream made by heating milk

        □ Single cream, double cream and whipping cream

        □ *Smetana*, Central and Eastern European variety of sour cream

    o Cultured milk resembling buttermilk, but uses different yeast and bacterial cultures

    o Kefir, fermented milk drink from the Northern Caucasus

    o *Kumis/Airag*, slightly fermented mares' milk popular in Central Asia

    o Powdered milk (or milk powder), produced by removing the water from (usually skim) milk

- ☐ Whole milk products

- ☐ Buttermilk products

- ☐ Skim milk

- ☐ Whey products

- ☐ High milk-fat and nutritional products (for infant formulas)

- ☐ Cultured and confectionery products

  - o Condensed milk, milk which has been concentrated by evaporation, with sugar added for reduced process time and longer life in an opened can

  - o *Khoa*, milk which has been completely concentrated by evaporation, used in Indian cuisine including gulab jamun, peda, etc.)

  - o Evaporated milk, (less concentrated than condensed) milk without added sugar

  - o Ricotta, acidified whey, reduced in volume

  - o Infant formula, dried milk powder with specific additives for feeding human infants

  - o Baked milk, a variety of boiled milk that has been particularly popular in Russia

- Butter, mostly milk fat, produced by churning cream

  - o Buttermilk, the liquid left over after producing butter from cream, often dried as livestock feed

  - o *Ghee*, clarified butter, by gentle heating of butter and removal of the solid matter

  - o *Smen*, a fermented, clarified butter used in Moroccan cooking

  - o Anhydrous milkfat (clarified butter)

- Cheese, produced by coagulating milk, separating from whey and letting it ripen, generally with bacteria and sometimes also with certain molds

  - o Curds, the soft, curdled part of milk (or skim milk) used to make cheese

  - o *Paneer*

  - o Whey, the liquid drained from curds and used for further processing or as a livestock feed

  - o Cottage cheese

  - o Quark

  - o Cream cheese, produced by the addition of cream to milk and then curdled to form a rich curd or cheese

  - o *Fromage frais*

- Casein are

    o Caseinates, sodium or calcium salts of casein

    o Milk protein concentrates and isolates

    o Whey protein concentrates and isolates, reduced lactose whey

    o Hydrolysates, milk treated with proteolytic enzymes to alter functionality

    o Mineral concentrates, byproduct of demineralizing whey

- Yogurt, milk fermented by *Streptococcus salivarius* ssp. *thermophilus* and *Lactobacillus delbrueckii* ssp. *bulgaricus* sometimes with additional bacteria, such as *Lactobacillus acidophilus*

    o *Ayran*

    o *Lassi*, Indian subcontinent

    o *Leben*

- Clabber, milk naturally fermented to a yogurt-like state

- Gelato, slowly frozen milk and water, lesser fat than ice cream

- Ice cream, slowly frozen cream, milk, flavors and emulsifying additives (dairy ice cream)

    o Ice milk, low-fat version of ice cream

    o Frozen custard

    o Frozen yogurt, yogurt with emulsifiers

- Other

    o *Viili*

    o *Kajmak*

    o *Filmjölk*

    o *Piimä*

    o *Vla*

    o *Dulce de leche*

    o *Skyr*

    o Junket, milk solidified with rennet

Health

Dairy products can cause health issues for individuals who have lactose intolerance or a milk allergy.

Additionally dairy products including cheese, ice cream, milk, butter, and yogurt can con-tribute significant amounts of cholesterol and saturated fat to the diet. Diets high in fat and especially in saturated fat can increase the risk of heart disease and can cause other serious health problems. However, it has been shown that there is no connection between dairy con-sumption (excluding butter) and cardiovascular disease, even though dairy tends to be higher in saturated fats.

There is no excess cardiovascular risk with dietary calcium intake but calcium supplements are associated with a higher risk of coronary artery calcification. Anderson JJ, Kruszka B, Delaney JA, et al. Calcium intake from diet and supplements and the risk of coronary artery calcification and its progression among older adults: 10-year follow-up of the Multi-Ethnic Study of Atherosclerosis (MESA). J Am Heart Assoc 20161; DOI:10.1161/jaha.116.003815

Consumption Patterns Worldwide

Rates of dairy consumption vary widely worldwide. High-consumption countries consume over 150 kg per capita per year: Argentina, Armenia, Australia, Costa Rica, Europe, Israel, Kyrgyzstan, North America and Pakistan. Medium-consumption countries consume 30 to 150 kg per capita per year: India, Iran, Japan, Kenya, Mexico, Mongolia, New Zealand, North and Southern Africa, most of the Middle East, and most of Latin America and the Caribbean. Low-consumption coun-tries consume under 30 kg per capita per year: Senegal, most of Central Africa, and most of East and Southeast Asia.

Avoidance

Some groups avoid dairy products for non-health related reasons:

- Religious - Some religions restrict or do not allow for the consumption of dairy products. For example, some scholars of Jainism advocate not consuming any dairy products be-cause dairy is perceived to involve violence against cows. Strict Judaism requires that meat and dairy products not be served at the same meal, served or cooked in the same utensils, or stored together, as prescribed in Deuteronomy 14:21.

- Ethical - Veganism is the avoidance of all animal products, including dairy products, most often due to the ethics regarding how dairy products are produced. The ethical reasons for avoiding dairy include how dairy is produced, how the animals are handled, and the envi-ronmental effect of dairy production.

Bulk Tank

In dairy farming a bulk milk cooling tank is a large storage tank for cooling and holding milk at a cold temperature until it can be picked up by a milk hauler. The bulk milk cooling tank is an im-portant piece of dairy farm equipment. It is usually made of stainless steel and used every day to store the raw milk on the farm in good condition. It must be cleaned after each milk collection. The milk cooling tank can be the property of the farmer or be rented from a dairy plant.

Bulk Tank Types

Different types of milk cooling tanks

Raw milk producers have a choice of either open (from 150 to 3000 litres) or closed (from 1000 to 10000 litres) tanks. The cost can vary considerably, depending on manufacturing norms and whether a new or second hand tank is purchased.

Milk silos (10,000 litres and plus) are suitable for the very large producer. These are designed to be installed outside and adjacent to the dairy, all controls and the milk outlet pipe being situated in the dairy.

Tank Construction

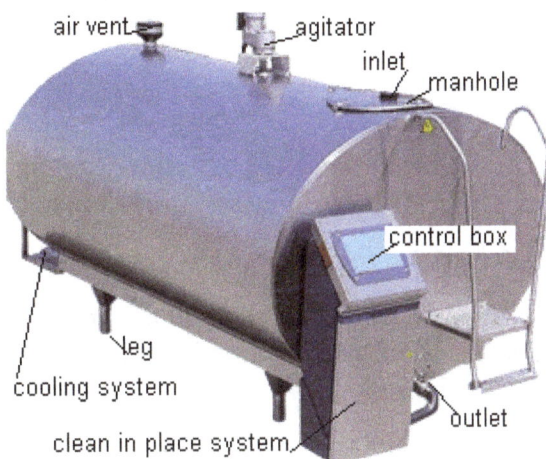

Bulk milk cooling tank description

A milk cooling tank, also known as a bulk tank or milk cooler, consists of an inner and an outer tank, both made of high quality stainless steel.

The space between the outer tank and the inner tank is isolated with polyurethane foam. In case of a power failure with an outside temperature of 30°C, the content of the tank will warm up only 1°C in 24 hours.

To facilitate an adequate and rapid cooling of the entire content of a tank, every tank is equipped with at least one agitator. Stirring the milk ensures that all milk inside the tank is of the same temperature and that the milk stays homogeneous.

On top of every closed milk cooling tank is a manhole of about 40 centimetres diameter. This enables thorough cleaning and inspection of the inner tank if necessary. The manhole is covered by a lid and sealed watertight with a rubber ring. Also on top are 2 or 3 small inlets. One is covered with an air-vent, the other(s) can be used to pump milk into the tank.

A milk cooling tank usually stands on 4, 6, or 8 adjustable legs. The built-in tilt of the inner tank ensures that even the last drop of milk will eventually flow to the outlet.

At the bottom, every milk cooling tank has a threaded outlet, usually including a valve.

All tanks have a thermometer, allowing for immediate inspection of the inner temperature.

Most tanks include an automatic cleaning system. Using hot and cold water, an acid and/or alkaline cleaning fluid, a pump and a spray lance will clean the inner tank, ensuring an hygienic inner environment each time the tank is emptied.

Almost every tank has a control box. It manages the cooling process by use of a thermostat. The user can turn the system on and off, allow for extra and immediate stirring, start the cleaning routine, and reset the entire system in case of a failure.

New and bigger milk cooling tanks are now being equipped with monitoring and alarm systems. These systems guard temperature of the milk inside the tank, check the functioning of the agitator, the cooling unit and temperature of the cleaning water. In case of malfunctioning of any of these functions, the alarm will activate. The monitoring system will also keep a record of the temperature and of all malfunctions for a given period.

Bulk Tank Manufacturing Norms

Norms define among other criteria: insulation, milk agitation, cooling power required, variations in milk quantity measurement, calibration, ... Some are more demanding than others.

- ISO standard 5708 (*Refrigerated bulk milk tanks*), published in 1983
- European standard EN 13732 (*Food processing machinery – Bulk milk coolers on farms – Requirements for construction, performance, suitability for use, safety and hygiene*), published in 2003, updated in 2009
- Northern American sanitary standard 3A 13-11 (*3-A Sanitary Standards for Farm Milk Cooling and Holding Tanks, 13-11*), effective July 23, 2012 www.3-a.org

Bulk Tank Outlet Standards

Swedish outlet (SMS 1145), German outlet (DIN 11851), English RJT (BS 4825), IDF (ISO 2853), tri-clamp (ISO 2852), Danish outlet (DS 722), ... can be found, not to mention different diameters. They vary from country to country. Non standard outlets make the milk collection process difficult, as the operator needs to adapt to each different standard/diameter.

Cooling Systems

There are two primary methods of cooling milk entering the bulk tank, each with its own advantages and disadvantages. The tank capacity and type will depend on herd size, calving pattern, frequency of milk collection, required milk quality, energy and water availability and future plans for development.

Direct Expansion

A bulk tank with direct expansion cooling has pipes or pillow plates carrying refrigerant which are welded directly to the exterior of the milk chamber. A layer of insulation covers the exterior of the milk tank and the cooling lines, with an exterior metal shell over the insulation.

Direct expansion cooling cannot run when the tank is empty or the inside walls of the tank would freeze. Instead, the tank is rapidly cooled as warm milk first enters the tank, and then the tank is cooled slowly just to maintain a low storage temperature. The rapid cooling during milking requires very large refrigeration compressors and condenser radiators to quickly expel heat from the milk, and is better suited for very large farming operations where three-phase electric power is available to operate the high-power cooling system.

Ice Bank

A bulk tank using an Ice Builder or Ice Bank immerses the bottom of the inner milk chamber in an open pool of water with copper tubes containing refrigerant suspended in the water. Between milkings, a small low-power cooling system slowly builds up a coating of ice around the copper tubes, and prevents icing of the pool over by continuously circulating the water in the pool. After the ice has achieved a thickness of 2-3 inches, the cooling system stops running.

During milking, the milk entering the tank is primarily cooled by circulating the water in the pool around the walls of the inner milk chamber, and the melting of the ice. After the ice has melted sufficiently the cooling system restarts to assist the ice bank and restart the ice building.

Ice bank bulk tanks are better suited for small family farm operations where only single-phase electric power is available, and high-power cooling systems would be either too expensive or difficult to install.

Milk Pre-cooling

For energy savings and quality reasons it is advisable to pre-cool the milk before it enters the tank using a plate or a tube cooler (shell and tube heat exchanger) supplied with chilled water from the well water, the ice builder or the condensing unit. The quicker milk is cooled after leaving the cow the better. This system achieves most of the cooling before the milk enters the tank, so that chilled milk, rather than warm milk, is being added to the already cooled milk in the tank.

Cooling Temperature

Generic temperature for milk storage is 3 to 4°C. For raw milk cheese manufacturing, it would be advisable to keep the milk at 12°C, as milk characteristics will be kept in a better state.

The milk cooling tank is usually not completely filled at once. A 2 milking tank is designed to cool 50% of its capacity at once. A 4 milking tank is designed to cool 25% of its capacity at once, and a 6 milking tank is designed to cool 16.7% of its capacity at once.

The cooling performance depends on the number of milking it takes to completely fill the tank, the ambient temperature and the cooling time.

Bulk Tank Cleaning Systems

There are two primary methods of cleaning bulk tanks, via manual scrubbing or automatic washing. Both methods generally use four steps to clean the tank:

- prerinsing with water to wet the surface and rinse off remaining milk residue
- washing with hot soapy water
- rinsing with water to remove the soap
- final *sanitizing* rinse with an approved bulk tank sanitizer solution

Manual Scrubbing

Manual scrubbing requires the bulk tank to have large hinged covers that can be lifted open to permit easy access to the interior surfaces of the tank. It tends to be much more thorough than automatic methods since it permits the tank to be carefully inspected during the washing process. If the tank is not found to be cleaned well enough, a troublesome area can be given additional cleansing attention.

Manual Scrubbing Limitations

This job is difficult to perform for very large tanks, and becomes more difficult as the overall cross-section or diameter of the tank increases, requiring either a longer brush or a raised work platform around the tank to lift the cleaning worker to reach over the side of a tall tank.

Automatic Washing

Automatic bulk tank washing and are normally activated by the milk collection truck driver after each milk collection. The cleaning system operates similar to a consumer dishwasher and consists of one or more free-spinning high-pressure spray nozzles with tangential jets, with the spray nozzle mounted on the end of a flexible whip suspended down into the center of the interior. As the cleaning solution sprays out of the jet, the force of the expelled water causes the jet to spin around and the whip to wildly swing back and forth, spraying the cleaning solution randomly all over the interior of the tank.

Automatic Washing Limitations

Because no physical scrubbing occurs with automatic wash systems, the cleanser relies on surfactants and detergents to dissolve the fats left on the interior of the tank by the cream in the milk. However, this is not sufficient to remove milkstone buildup, and the tank may need to be washed occasionally with milkstone remover to remove this scale buildup that can harbor bacteria and contaminants.

Automatic scrubbing only cleans the interior of the tank. It is not capable of cleaning the exterior of the tank, and it does not do a good job of washing around the cover seals. While it is possible to just clean the interior and call it good enough, it does not provide the maximum sanitation of manually washing down the exterior of the tank following or during the automatic wash process. Also, some components that contact the milk such as the drain valve cannot be properly cleaned automatically without disassembling the valve and retaining washer and directly scrubbing in soapy water.

Operating Costs

Substantial reductions in running costs can be made when an ice builder is used in conjunction with off-peak electricity. Pre-cooling milk using a plate or a tube cooler supplied with mains or well water can also reduce costs and add to the cooling capacity of the tank.

Bulk tank condenser units, which are not an integral part of the tank, should be fitted in an adjacent, suitable and well ventilated place.

If at all possible, condenser units should not be fitted on a wall facing the sun. They should be installed in a way which allows them to draw in and discharge adequate quantities of air for efficient operation.

Bulk tank should be easily accessible by large bulk collection tankers and positioned so that the tanker approaches can be kept clean and free from cow traffic at all times.

Although tanks have been calibrated when first installed, bulk tank miscalibration is not uncommon and in some cases it can result in significant loss of income. Milk tanks calibrated on the low side, can cheat raw milk producers by up to 22 litres on each shipment. It is therefore advisable to re-calibrate a bulk tank.

Other Usage of Bulk Tanks

Stainless steel bulk tanks are also used to heat or cool a fluid or simply to keep it isolated and warm/cold. Because of the hygienical finishing of the inner and outer side of the tanks, almost any fluid can be stored: water, fruit juices, honey, wine, beer, ink, paint, cosmetics, aromatic food-additives, bacterial cultures, cleansers, oil, or blood.

References

- Sisney, Jason; Garosi, Justin. "California is the Leading Farm State". Legislative Analyst's Office. Retrieved 28 May 2016.

- U.S. Department of Agriculture, National Agriculture Statistics Service (April 2015). "Livestock Slaughter Annual Summary" (PDF). Retrieved 2015-11-24.

- "Purebred Dairy Cattle Association". Purebred Dairy Cattle Association. Purebred Dairy Cattle Association. Retrieved 17 September 2014.

- Collier RJ, Bauman DE. Update on human health concerns of recombinant bovine somatotropin use in dairy cows. J Anim Sci. 2014 Apr;92(4):1800-7. doi: 10.2527/jas.2013-7383. PMID 24663163.

- Collier, R. J., & Bauman, D. E. (2014). Update on human health concerns of recombinant bovine somatotropin use in dairy cows. Journal of Animal Science, 92(4), 1800-1807. 10.2527/jas2013-7383.

- McLean, Amy. "Donkey milk for human health?". Tri-State Livestock News. Swift Communications, Inc. Retrieved 28 December 2013.

4

Poultry Farming

Raising domesticated birds is known as poultry farming. These birds include ducks, chickens and geese and are used for the purpose of farming eggs and meat. Poultry, free range, yarding, battery cage, furnished cages and broiler industry are some of the aspects of poultry farming. The aspects elucidated in this section help the reader in developing a better understanding of poultry farming.

Poultry Farming

Poultry farming is the raising of domesticated birds such as chickens, ducks, turkeys and geese for the purpose of farming meat or eggs for food. Poultry are farmed in great numbers with chickens being the most numerous. More than 50 billion chickens are raised annually as a source of food, for both their meat and their eggs. Chickens raised for eggs are usually called layers while chickens raised for meat are often called broilers. In the US, the national organization overseeing poultry production is the Food and Drug Administration (FDA). In the UK, the national organisation is the Department for Environment, Food and Rural Affairs (Defra).

Intensive and Alternative

According to the researchers and scientists, 74% of the world's poultry meat, and 68 percent of eggs are produced in ways that are described as 'intensive'. One alternative to intensive poultry farming is free-range farming using lower stocking densities. Poultry producers routinely use nationally approved medications, such as antibiotics, in feed or drinking water, to treat disease or to prevent disease outbreaks. Some FDA-approved medications are also approved for improved feed utilization.

Egg-laying Chickens – husbandry Systems

Commercial hens usually begin laying eggs at 16–20 weeks of age, although production gradually declines soon after from approximately 25 weeks of age. This means that in many countries, by approximately 72 weeks of age, flocks are considered economically unviable and are slaughtered after approximately 12 months of egg production, although chickens will naturally live for 6 or more years. In some countries, hens are force moulted to re-invigorate egg-laying.

Environmental conditions are often automatically controlled in egg-laying systems. For example, the duration of the light phase is initially increased to prompt the beginning of egg-laying at 16–20 weeks of age and then mimics summer daylength which stimulates the hens to continue laying eggs all year round; normally, egg production occurs only in the warmer months. Some commercial breeds of hen can produce over 300 eggs a year.

Free-range

Commercial free range hens

Free range chickens being fed outdoors

Free-range poultry farming allows chickens to roam freely for a period of the day, although they are usually confined in sheds at night to protect them from predators or kept indoors if the weather is particularly bad. In the UK, the Department for Environment, Food and Rural Affairs (Defra) states that a free-range chicken must have day-time access to open-air runs during at least half of its life. Unlike in the United States, this definition also applies to free-range egg laying hens. The European Union regulates marketing standards for egg farming which specifies a minimum condition for free-range eggs that "hens have continuous daytime access to open-air runs, except in the case of temporary restrictions imposed by veterinary authorities". The RSPCA "Welfare standards for laying hens and pullets" indicates that the stocking rate must not exceed 1,000 birds per hectare (10 m² per hen) of range available and a minimum area of overhead shade/shelter of 8 m² per 1,000 hens must be provided.

Free-range farming of egg-laying hens is increasing its share of the market. Defra figures indicate that 45% of eggs produced in the UK throughout 2010 were free-range, 5% were produced in barn systems and 50% from cages. This compares with 41% being free-range in 2009.

Suitable land requires adequate drainage to minimise worms and coccidial oocysts, suitable protection from prevailing winds, good ventilation, access and protection from predators. Excess heat, cold or damp can have a harmful effect on the animals and their productivity. Free-range farmers have less control than farmers using cages in what food their chickens eat, which can lead to unreliable productivity, though supplementary feeding reduces this uncertainty. In some farms, the manure from free-range poultry can be used to benefit crops.

The benefits of free-range poultry farming for laying hens include opportunities for natural be-haviours such as pecking, scratching, foraging and exercise outdoors.

Both intensive and free-range farming have animal welfare concerns. Cannibalism, feather peck-ing and vent pecking can be common, prompting some farmers to use beak trimming as a preven-tative measure, although reducing stocking rates would eliminate these problems. Diseases can be common and the animals are vulnerable to predators. Barn systems have been found to have the worst bird welfare. In South-East Asia, a lack of disease control in free range farming has been associated with outbreaks of Avian influenza.

Organic

In organic egg-laying systems, chickens are also free-range. Organic systems are based upon re-strictions on the routine use of synthetic yolk colourants, in-feed or in-water medications, other food additives and synthetic amino acids, and a lower stocking density and smaller group sizes. The Soil Association standards used to certify organic flocks in the UK, indicate a maximum out-doors stocking density of 1,000 birds per hectare and a maximum of 2,000 hens in each poultry house. In the UK, organic laying hens are not routinely beak-trimmed.

Yarding

While often confused with free-range farming, yarding is actually a separate method of poultry culture by which chickens and cows are raised together. The distinction is that free-range poultry are either totally unfenced, or the fence is so distant that it has little influence on their freedom of movement. Yarding is common technique used by small farms in the Northeastern US. The birds are released daily from hutches or coops. The hens usually lay eggs either on the floor of the coop or in baskets if provided by the farmer. This husbandry technique can be complicated if used with roosters, mostly because of aggressive behavior.

Battery Cage

Bank of cages

The majority of hens in many countries are reared in battery cages, although the European Union

Council Directive 1999/74/EC has banned the conventional battery cage in EU states from January 2012. These are small cages, usually made of metal in modern systems, housing 3 to 8 hens. The walls are made of either solid metal or mesh, and the floor is sloped wire mesh to allow the faeces to drop through and eggs to roll onto an egg-collecting conveyor belt. Water is usually provided by overhead nipple systems, and food in a trough along the front of the cage replenished at regular intervals by a mechanical chain.

The cages are arranged in long rows as multiple tiers, often with cages back-to-back (hence the term 'battery cage'). Within a single shed, there may be several floors containing battery cages meaning that a single shed may contain many tens of thousands of hens. Light intensity is often kept low (e.g. 10 lux) to reduce feather pecking and vent pecking. Benefits of battery cages include easier care for the birds, floor eggs which are expensive to collect are eliminated, eggs are cleaner, capture at the end of lay is expedited, generally less feed is required to produce eggs, broodiness is eliminated, more hens may be housed in a given house floor space, internal parasites are more easily treated, and labor requirements are generally much reduced.

In farms using cages for egg production, there are more birds per unit area; this allows for greater productivity and lower food costs. Floor space ranges upwards from 300 cm² per hen. EU standards in 2003 called for at least 550 cm² per hen. In the US, the current recommendation by the United Egg Producers is 67 to 86 in² (430 to 560 cm²) per bird. The space available to battery hens has often been described as less than the size of a piece of A4 paper. Animal welfare scientists have been critical of battery cages because they do not provide hens with sufficient space to stand, walk, flap their wings, perch, or make a nest, and it is widely considered that hens suffer through boredom and frustration through being unable to perform these behaviours. This can lead to a wide range of abnormal behaviours, some of which are injurious to the hens or their cagemates.

Furnished Cage

In 1999, the European Union Council Directive 1999/74/EC banned conventional battery cages for laying hens throughout the European Union from January 1, 2012; they were banned previously in other countries including Switzerland. In response to these bans, development of prototype commercial furnished cage systems began in the 1980s. Furnished cages, sometimes called 'enriched' or 'modified' cages, are cages for egg laying hens which have been designed to overcome some of the welfare concerns of battery cages whilst retaining their economic and husbandry advantages, and also provide some of the welfare advantages of non-cage systems. Many design features of furnished cages have been incorporated because research in animal welfare science has shown them to be of benefit to the hens. In the UK, the Defra "Code for the Welfare of Laying Hens" states furnished cages should provide at least 750 cm² of cage area per hen, 600 cm² of which should be usable; the height of the cage other than that above the usable area should be at least 20 cm at every point and no cage should have a total area that is less than 2000 cm². In addition, furnished cages should provide a nest, litter such that pecking and scratching are possible, appropriate perches allowing at least 15 cm per hen, a claw-shortening device, and a feed trough which may be used without restriction providing 12 cm per hen.

Modern egg laying breeds often suffer from osteoporosis which results in the chicken's skeletal system being weakened. During egg production, large amounts of calcium are transferred from bones to create egg-shell. Although dietary calcium levels are adequate, absorption of dietary calcium is not always sufficient, given the intensity of production, to fully replenish bone calcium. This can lead to increases in bone breakages, particularly when the hens are being removed from cages at the end of laying.

Countries such as Austria, Belgium or Germany are planning to ban furnished cages until 2025 additionally to the already banned conventional cages.

Meat-producing Chickens – husbandry Systems

Broilers in a production house

Indoor Broilers

Meat chickens, commonly called broilers, are floor-raised on litter such as wood shavings, peanut shells, and rice hulls, indoors in climate-controlled housing. Under modern farming methods, meat chickens reared indoors reach slaughter weight at 5 to 9 weeks of age. The first week of chickens life they can grow 300 percent of their body size, a nine-week-old chicken can average over 9 pounds in body weight. At nine weeks a hen will average around 7 pounds and a rooster will weigh around 12 pounds, having a nine-pound average.

Broilers are not raised in cages. They are raised in large, open structures known as grow out houses. A farmer receives the birds from the hatchery at one day old. A grow out consist of 5 to 9 weeks according on how big the kill plant wants the chickens to be. These houses are equipped with mechanical systems to deliver feed and water to the birds. They have ventilation systems and heaters that function as needed. The floor of the house is covered with bedding material consisting of wood chips, rice hulls, or peanut shells. In some cases they can be grown over dry litter or compost. Because dry bedding helps maintain flock health, most growout houses have enclosed watering systems ("nipple drinkers") which reduce spillage.

Keeping birds inside a house protects them from predators such as hawks and foxes. Some houses are equipped with curtain walls, which can be rolled up in good weather to admit natural light and fresh air. Most growout houses built in recent years feature "tunnel ventilation," in which a bank of fans draws fresh air through the house.

Traditionally, a flock of broilers consist of about 20,000 birds in a growout house that measures 400/500 feet long and 40/50 feet wide, thus providing about eight-tenths of a square foot per bird. The Council for Agricultural Science and Technology (CAST) states that the minimum space is one-half square foot per bird. More modern houses are often larger and contain more birds, but the floor space allotment still meets the needs of the birds. The larger the bird is grown the fewer chickens are put in each house, to give the bigger bird more space per square foot.

Because broilers are relatively young and have not reached sexual maturity, they exhibit very little aggressive conduct.

Chicken feed consists primarily of corn and soybean meal with the addition of essential vitamins and minerals. No hormones or steroids are allowed in raising chickens.

Issues with Indoor Husbandry

In intensive broiler sheds, the air can become highly polluted with ammonia from the droppings. In this case a farmer must run more fans to bring in more clean fresh air. If not this can damage the chickens' eyes and respiratory systems and can cause painful burns on their legs (called hock burns) and blisters on their feet. Broilers bred for fast growth have a high rate of leg deformities because the large breast muscles causes distortions of the developing legs and pelvis, and the birds cannot support their increased body weight. In cases where the chickens become crippled and can't walk farmers have to go in and pull them out. Because they cannot move easily, the chickens are not able to adjust their environment to avoid heat, cold or dirt as they would in natural conditions. The added weight and overcrowding also puts a strain on their hearts and lungs and Ascites can develop. In the UK, up to 19 million broilers die in their sheds from heart failure each year. In the case of no ventilation due to power failure during a heat wave 20,000 chicken can die in a short period of time. In a good grow out a farmer should sell between 92 and 96 percent of their flock. With a 1.80 to a 2.0 feed conversion ratio. After the marking of birds the farmer must clean out and repair for another flock. A farmer should average 4 to 5 grow outs a year.

Indoor with Higher Welfare

Chickens are kept indoors but with more space (around 12 to 14 birds per square metre). They have a richer environment for example with natural light or straw bales that encourage foraging and perching. The chickens grow more slowly and live for up to two weeks longer than intensively farmed birds. The benefits of higher welfare indoor systems are the reduced growth rate, less crowding and more opportunities for natural behaviour.

Free-range Broilers

Turkeys on pasture at an organic farm

Free-range broilers are reared under similar conditions to free-range egg laying hens. The breeds grow more slowly than those used for indoor rearing and usually reach slaughter weight at approximately 8 weeks of age. In the EU, each chicken must have one square metre of outdoor space. The benefits of free-range poultry farming include opportunities for natural behaviours such as pecking, scratching, foraging and exercise outdoors. Because they grow slower and have opportunities for exercise, free-range broilers often have better leg and heart health.

Organic Broilers

Organic broiler chickens are reared under similar conditions to free-range broilers but with restrictions on the routine use of in-feed or in-water medications, other food additives and synthetic amino acids. The breeds used are slower growing, more traditional breeds and typically reach slaughter weight at around 12 weeks of age. They have a larger space allowance outside (at least 2 square metres and sometimes up to 10 square metres per bird). The Soil Association standards indicate a maximum outdoors stocking density of 2,500 birds per hectare and a maximum of 1,000 broilers per poultry house.

Issues

Humane Treatment

Battery cages

Chickens transported in a truck.

Animal welfare groups have frequently criticized the poultry industry for engaging in practices which they believe to be inhumane. Many animal rights advocates object to killing chickens for food, the "factory farm conditions" under which they are raised, methods of transport, and slaughter. Compassion Over Killing and other groups have repeatedly conducted undercover investigations at chicken farms and slaughterhouses which they allege confirm their claims of cruelty.

Conditions in chicken farms may be unsanitary, allowing the proliferation of diseases such as salmonella, E. coli and campylobacter. Chickens may be raised in very low light intensities, sometimes total darkness, to reduce injurious pecking. Concerns have been raised that companies growing single varieties of birds for eggs or meat are increasing their susceptibility to disease. Rough handling, crowded transport during various weather conditions and the failure of existing stunning systems to render the birds unconscious before slaughter, have also been cited as welfare concerns.

A common practice among hatcheries for egg-laying hens is the culling of newly hatched male chicks since they do not lay eggs and do not grow fast enough to be profitable for meat. There are plans to more ethically destroy the eggs before the chicks are hatched by "in-ovo" sex determination.

Beak Trimming

Laying hens are routinely beak-trimmed at 1 day of age to reduce the damaging effects of aggression, feather pecking and cannibalism. Scientific studies have shown that beak trimming is likely to cause both acute and chronic pain.

The beak is a complex, functional organ with an extensive nervous supply including nociceptors that sense pain and noxious stimuli. These would almost certainly be stimulated during beak trimming, indicating strongly that acute pain would be experienced. Behavioural evidence of pain after beak trimming in layer hen chicks has been based on the observed reduction in pecking behavior, reduced activity and social behavior, and increased sleep duration. Severe beak trimming, or beak trimming birds at an older age, may cause chronic pain. Following beak trimming of older or adult hens, the nociceptors in the beak stump show abnormal patterns of neural discharge, which indicate acute pain.

Neuromas, tangled masses of swollen regenerating axon sprouts, are found in the healed stumps of birds beak trimmed at 5 weeks of age or older and in severely beak trimmed birds. Neuromas have been associated with phantom pain in human amputees and have therefore been linked to chronic pain in beak trimmed birds. If beak trimming is severe because of improper procedure or done in older birds, the neuromas will persist which suggests that beak trimmed older birds experience chronic pain, although this has been debated.

Beak-trimmed chicks will initially peck less than non-trimmed chickens, which animal behavioralist Temple Grandin attributes to guarding against pain. The animal rights activist, Peter Singer, claims this procedure is bad because beaks are sensitive, and the usual practice of trimming them without anaesthesia is considered inhumane by some. Some within the chicken industry claim that beak-trimming is not painful whereas others argue that the procedure causes chronic pain and discomfort, and decreases the ability to eat or drink.

Antibiotics

Overview of Antibiotic Use in Poultry Antibiotics have been used in poultry farming in mass quantities since 1951, when the Food and Drug Administration (FDA) approved their use. Three years prior to the FDA's approval, scientists were investigating a phenomena in which chickens who were rooting through bacteria-rich manure were displaying signs of greater health than those who did not. Through testing, it was discovered that chickens who were fed a variety of vitamin B12 manufactured with the residue of a certain antibiotic grew 50 percent faster than those chickens who were fed B12 manufactured from a different source. Further testing confirmed that use of antibiotics did improve the health of the chickens, resulting in the chickens laying more eggs and experiencing lower mortality rates and less illness. Upon this discovery, farmers transitioned from expensive animal proteins to comparatively inexpensive antibiotics and B12. Chickens were now reaching their market weight at a much faster rate and at a lower cost. With a growing population and greater demand on the farmers, antibiotics appeared to be an ideal and cost-effective way to increase the output of poultry. Since this discovery, antibiotics have been routinely used in poultry production, but more recently have been the topic of debate secondary to the fear of bacterial antibiotic resistance.

Emerging Threats: Antibiotic Resistance The Centers for Disease Control (CDC), has identified the emergence of antibiotic resistance as a national threat. The concern over antibiotic use in livestock arises from the necessity antibiotics have in keeping populations disease-free. As of 2016, over 70 percent of FDA approved antibiotics are utilized in modern, high production poultry farms to prevent, control, and treat disease. The FDA released a report in 2009 estimating that 29 million pounds of antibiotics had been used in livestock in that year alone. However, surveillance of consumer exposure to antibiotics through poultry consumption is limited. More specifically in 2012, the FDA speculated the most significant public health threat in regard to antimicrobial use in animals is the exposure of antimicrobial resistant bacteria to humans. These statements are challenged by the American meat industry lobbyists that antibiotics are used responsibly and judiciously in order to ensure effectiveness.

Consumer Health effects Consumers are exposed to antibiotic resistance through consumption of poultry products that have prior exposure to resistant strains. In poultry husbandry, the practice of using medically important antibiotics can select for resistant strains of bacteria, which are then transferred to consumers through poultry meat and eggs. The CDC acknowledges this transferal pathway in their 2013 report of Antibiotic Resistant Threats in the United States. The annual rate of foodborne illness in the United States is one in six. For the 48 million individuals affected, antibiotics play a critical role in thwarting mortality rates. In a literature review conducted by the Review of Antimicrobial Resistance 100 out of 139 studies found evidence of a link between antibiotic use in animals and antibiotic resistance in consumers.

When a gram-negative bacterial infection is suspected in a patient, one of the first-line options for treatment is in the fluoroquinolone family. This, along with penicillin, is one of the first families of antibiotics utilized in the broiler industry. If this first-line treatment is not successful, a stronger class of antibiotics is typically used, however, there is a limitation on how many classes are available, as well as which medications are available on hospital formularies. There is also more drug toxicity affiliated with second and third line antibiotic options. This is one example why it is critical to keep as many first line antibiotic options available for human use.

Other issues are associated with duration and complexity of infection. On average, treatment for non-resistant bacteria is administered 11.5 hours after diagnosis, and treatment for resistant bacteria is administered 72 hours after diagnosis. This is a reflection of the additional threat of prolonged incubation, leading to greater potential for systemic disease, with higher morbidity and mortality associated with opportunities for complications, and prolonged treatment time. For example, of the two million people affected by resistant infections a year, 23,000 will die. Severity in mortality is coupled when exposed to high risk populations such as immunocompromised and elderly individuals in hospital and nursing home settings.

History of US Federal policy on antibiotic use in livestock

- 1940s – Beginning of utilization of antibiotics in livestock feed

- 1951 - Antibiotics first FDA approved for use in poultry. Approved uses included production (growth enhancement), treatment, control, or prevention of animal disease. Antibiotics were also available for purchase over the counter at that time.

- 1970 - FDA task force publication proposes limitations of utilizing antibiotics in livestock feed that are also used in humans.

- 1975 - Secondary to this publication, drug sponsors are required to submit studies demonstrating the antibiotics did not harm human health

- 1976 - Stuart Levy study demonstrating tetracycline resistant E. coli moving to consumers

- 1977 - FDA proposal to remove penicillin and tetracycline in subtherapeutic doses, however, request by Congress for further studies to be conducted.

- 1980 - National Academy of Science recruited by the FDA to conduct further studies, specifically for penicillins and tetracyclines. Conclusion from these studies indicated no sufficient evidence to ban these antibiotics.

- 1980s-early 2000s - Further studies continued, supported by the FDA

- 2003 - FDA issued guidance to pharmaceuticals for an approval process utilizing new antibiotics in animal feed. For antibiotics already in use, the FDA would have to withdraw approval for each individual medication.

- 2005 - Enrofloxacin, an already utilized antibiotic, was removed from poultry production. This took 5 years to accomplish.

- 2010 - FDA first draft of "voluntary" limitations of medically important antibiotics in livestock, and requirement of veterinarian oversight, which would later become "Guidance for Industry #209."

- 2011 - FDA removed original request from 1977 to remove penicillins and tetracyclines in feed.

- 2012 - FDA finalized "Guidance for Industry #209," which was implemented under the Veterinary Feed Directives. These guidelines were issued to pharmaceuticals.

- 2013 - FDA issues "Guidance for Industry #213," which provided additional information to pharmaceuticals for recommendations from #209.

- 2014 - All 26 pharmaceutical companies producing antibiotics used in livestock feed agreed to the FDA guidelines in #213. Gave total of 3 years to make all recommended changes.

Current Federal Regulators National Antimicrobial Resistance Monitoring System's (NARMS) Enteric Bacteria program - Established in 1996, and represents a collaboration between the USDA,

FDA, and CDC. Its purpose is to organize these organizations into a drug monitoring program for antibiotics utilized in animal feed withe goal of maintaining their medical efficacy. There are three branches which oversee humans, retail meat, and food animals.

- USDA - Operating under the Food Safety and Inspection Service (FSIS). Main role is in charge of testing imported and domestic meat for antimicrobial resistant bacteria. If a 'residue violation' found, they may condemn the product. Regardless, funding and resources are not available for outbreak investigations at farms or ranches.

- FDA - Operating under the Center for Veterinary Medicine (CVM). Works with CDC to monitor retail meat.

- CDC - Monitors human samples.

Vertical Integration This is the current business structure utilized almost universally in the broiler, or chicken bred for meat, industry. This also began in the 1940s when antibiotics began to be utilized in livestock feed. Perdue is credited as the pioneer of this structure. The basis is centralization of production. 'Integrators' control cost, policy, and are the decision makers of production. They decide feed formulations, choice of antibiotic administration, and cover those costs in addition to veterinary services. They also own the poultry that is grown. Farmers are labeled as 'Growers' or 'Operators'. They own the land and buildings where the poultry is grown, and are essentially caretakers for the poultry growth to the Integrators. The benefit for Growers in this business structure is they are guaranteed payment from the Integrators, which is compensated in weight gained by each flock. Due to this structure, about 90% of broilers are raised within 60 miles of the processing plant. Integrators are large poultry companies such as Perdue, Tyson, Pilgrim's Pride, Koch Foods, etc. There are about 20 of these companies in the U.S. that control 96% of all broilers produced in 2011.

Regulatory Surveys There are two main surveys distributed to farmers by the federal government to aid in various regulations of the agricultural industry. They are the Agricultural and Resource Management Survey (ARMS) and the National Animal Health Monitoring Survey (NAHMS).

Agricultural and Resource Management Survey (ARMS) - Ran by the USDA's Economic Research Service (ERS) and National Agricultural Statistics Service (NASS). The main focus is finances of farming, production practices, and resource use. Seventeen total states are sampled every 5–6 years per livestock type, with the most recent surveys distributed to broiler farmers in 2006 and 2011. There was one question about utilization of antibiotics in poultry food or water, excluding use for illness treatment.

Antibiotic Resistant Outbreaks from Poultry Meat In order to minimize and prevent any residues of antibiotics in chicken meat, any chickens given antibiotics are required to have a "withdrawal" period before they can be slaughtered. Samples of poultry at slaughter are randomly tested by the FSIS, and show a very low percentage of residue violations. Although violations are minimal, these small amounts of antibiotics have still contributed to antibiotic resistant outbreaks in the U.S. There are five infectious agents that account for 90% of foodborne related deaths. Three consistently found in poultry are: Salmonella, Campylobacter, and Escherichia coli.

- 2014: Outbreak of Salmonella in 634 people across 29 states (38% hospitalized) from eating chicken from Foster Farms that was sold at Costco. 44/68 tested isolates were resistant to at least 1 drug (65%), and 4 of 5 chicken samples tested were drug resistant (80%).

- 2015: Outbreak of Salmonella in 15 people in 7 states (4 hospitalized) from eating frozen stuffed chicken produced by Barber Foods.

Limitations & Challenges One obstacle to gathering more comprehensive data on the use of antibiotics in feed is the majority of the poultry industry utilizes vertical integration. As a consequence, farmers are often unaware of what components go into the feed, including whether or not antibiotics are used. Also in antibiotic usage in general, there are criteria to define bacterial resistance to specific antibiotics, however, there are no standards to divide the bacteria into resistant and susceptible categories based on antibiotics utilized.

The poultry industry also plays a large part in the United States economy, both in domestic purchasing and through international demand. The USDA reports that the U.S. is the "world's largest producer and second largest exporter of poultry meat." In 2010, the U.S. produced 36.9 billion pounds of broiler meat and exported 6.8 billion pounds of broiler meat. This equates to an estimated retail value of 45 billion dollars in 2010.

Both the agricultural and pharmaceutical industries have been lobbying against legislation that seeks to quell non-therapeutic antibiotic use in livestock since the first introduction of such legislation in Congress in the 1970's. Despite scientific evidence suggesting a strong association between antibiotic use in poultry and other livestock, agribusiness lobbies such as The National Chicken Council argue that there is not sufficient evidence to purport that there is a measurable impact to humans and shifts the blame of the problem of antibiotic resistance to overprescribing in the field of medicine.

With antibiotic restrictions, integrators will bare the immediate costs of these changes, and would likely result in modified finances and contracts with growers. Also, public health agencies may not have adequate scientific evidence for making appropriate decisions for better public health outcomes, secondary to lack of research funds. As a reference, the US spends about $101 billion per year for both governmental and biomedical industrial research, which is only 5% of total health expenditures.

Solutions Several policies have been proposed to improve data collection and transparency in livestock production. For example, the 2013 Delivering Antimicrobial Transparency in Animals (DATA) Act proposed the enactment of policies to acquire more accurate documentation of antibiotic use in growth promotion by farmers, drug manufacturers, and the FDA. Also, the Preservation of Antibiotics for Medical Treatment Act (PAMTA) was enacted to eliminate the use of medically important antibiotics in livestock. In 2015, the Preventing Antibiotic Resistance Act (PARA) was passed with two components: requirement of drug companies to provide evidence that antibiotics that are approved for use in poultry, and that meat production does not add to the growing threat of antibiotic resistance in humans. Antimicrobial Stewardship Programs (ASPs) serve as an example of systematic monitoring and analysis of data via interdisciplinary and multi-sectoral collaboration.

Performing quality improvement in the process of livestock production is another focus. Some alternative methods include "improving hygiene, using enzymes, probiotics, prebiotics, and acids to improve health and utilizing bacteriocins, antimicrobial peptides, and bacteriophages as substitutes for antibiotics." Adaptations of methods by other countries is an addtional focus. For example, the use of antibiotics in feed was banned in Sweden in 1985 with no compensatory increase in antibiotic usage in other sectors of production, proving that a ban can be successfully administered without unintended impacts on other categories.

Major producers in the poultry industry have also begun to make strides towards change, largely due to public concern over the widespread use of antibiotics in poultry. Some producers have started eliminating the use of antibiotics in order to produce and market chickens that may legally be labeled "antibiotic free". In 2007, Perdue began phasing out all medically important antibiotics from its feed and hatcheries and began selling poultry products labeled "no antibiotics ever" under the Harvestland brand. Consumer response was positive and in 2014 Perdue also began phasing out ionophores from its hatchery and began using the "antibiotic free" labels on its Harvestland, Simply Smart and Perfect Portions products.

Impacts of Change As Guidance for Industry #213 has been voluntarily accepted, it will be a violation of the Federal Food, Drug, and Cosmetic Act to use antibiotics in livestock production for non-therapeutic purposes. However, as there is now a requirement for veterinary oversight and approval for antibiotics use, there is leeway in the interpretation of non-therapeutic purposes dependent on the situation. For example, per the FDA, "a veterinarian may determine, based on the client's production practices and history, that weaned beef calves arriving at a feedlot in bad weather after a lengthy transport are at risk to develop bacterial respiratory infection. In this case, the veterinarian might choose to preventively treat these calves with an antimicrobial approved for prevention of that bacterial infection."

The FDA is not trying to regulate all antimicrobials at this time - only those antibiotics which are considered "medically important." For example, bacitracin, a common antibiotic found in over the counter antibiotic ointments, is not classified as "medically important." Also, ionophores, which are not apart of human medicine but given for improving the health of livestock, are also not included in this regulation.

Arsenic

Poultry feed can also include roxarsone or nitarsone, arsenical antimicrobial drugs that also promote growth. Roxarsone was used as a broiler starter by about 70% of the broiler growers between 1995 and 2000. The drugs have generated controversy because it contains arsenic, which is highly toxic to humans. This arsenic could be transmitted through run-off from the poultry yards. A 2004 study by the U.S. magazine Consumer Reports reported "no detectable arsenic in our samples of muscle" but found "A few of our chicken-liver samples has an amount that according to EPA standards could cause neurological problems in a child who ate 2 ounces of cooked liver per week or in an adult who ate 5.5 ounces per week." The U.S. Food and Drug Administration (FDA), however, is the organization responsible for the regulation of foods in America, and all samples tested were "far less than the... amount allowed in a food product."

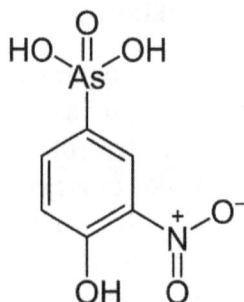

Roxarsone, a controversial arsenic compound used as a nutritional supplement for chickens.

Growth Hormones

Hormone use in poultry production is illegal in the United States. Similarly, no chicken meat for sale in Australia is fed hormones. Several scientific studies have documented the fact that chickens grow rapidly because they are bred to do so, not because of growth hormones. A small producer of natural and organic chickens confirmed this assumption:

> Using hormones to boost egg production was a brief fad in the Forties, but was abandoned because it didn't work. Using hormones to produce soft-meated roasters lasted into the Fifties, but the improved growth rates of normal, untreated broilers made the practice irrelevant—the broilers got as big as anyone wanted without chemicals. The only hormone that was ever used in any quantity on poultry (DES) was banned in 1959, and everyone but a few die-hard farmers had given up hormones by then, anyway. Hormones are now illegal in poultry and eggs.

E. coli

According to Consumer Reports, "1.1 million or more Americans [are] sickened each year by under-cooked, tainted chicken." A USDA study discovered *E. coli (Biotype I)* in 99% of supermarket chicken, the result of chicken butchering not being a sterile process. However, the same study also shows that the strain of *E. coli* found was always a non-lethal form, and no chicken had any of the pathenogenic O157:H7 serotype. Many of these chickens, furthermore, had relatively low levels of contamination.

Feces tend to leak from the carcass until the evisceration stage, and the evisceration stage itself gives an opportunity for the interior of the carcass to receive intestinal bacteria. (The skin of the carcass does as well, but the skin presents a better barrier to bacteria and reaches higher temperatures during cooking.) Before 1950, this was contained largely by not eviscerating the carcass at the time of butchering, deferring this until the time of retail sale or in the home. This gave the intestinal bacteria less opportunity to colonize the edible meat. The development of the "ready-to-cook broiler" in the 1950s added convenience while introducing risk, under the assumption that end-to-end refrigeration and thorough cooking would provide adequate protection. *E. coli* can be killed by proper cooking times, but there is still some risk associated with it, and its near-ubiquity in commercially farmed chicken is troubling to some. Irradiation has been proposed as a means of sterilizing chicken meat after butchering.

The aerobic bacteria found in poultry housing can include not only *E. coli*, but *Staphylococcus, Pseudomona, Micrococcus* and others as well. These contaminants can contribute to dust that often cause issues with the respiratory systems of both the poultry and humans working in the environment. If bacterial levels in the poultry drinking water reach high levels, it can result in bacterial diarrhoea which can lead to blood poisoning should the bacteria spread from the damaged intestines.

Salmonella too can be stressful on poultry production. How it causes disease has been investigated in some detail.

Avian Influenza

There is also a risk that crowded conditions in chicken farms will allow avian influenza (bird flu) to spread quickly. A United Nations press release states: "Governments, local authorities and international agencies need to take a greatly increased role in combating the role of factory-farming, commerce in live poultry, and wildlife markets which provide ideal conditions for the virus to spread and mutate into a more dangerous form..."

Efficiency

Farming of chickens on an industrial scale relies largely on high protein feeds derived from soybeans; in the European Union the soybean dominates the protein supply for animal feed, and the poultry industry is the largest consumer of such feed. Two kilograms of grain must be fed to poultry to produce 1 kg of weight gain, much less than that required for pork or beef. However, for every gram of protein consumed, chickens yield only 0.33 g of edible protein.

Economic Factors

Changes in commodity prices for poultry feed have a direct effect on the cost of doing business in the poultry industry. For instance, a significant rise in the price of corn in the United States can put significant economic pressure on large industrial chicken farming operations.

Worker Health and Safety

Poultry workers experience substantially higher rates of illness and injury than manufacturing workers do on average.

Muscular Disorders

For the year 2013, there were an estimated 1.59 cases of occupation-related illness per 100 full time U.S. meat and poultry workers, compared to .36 for manufacturing workers overall. Injuries are associated with repetitive movements, awkward postures, and cold temperatures. High rates of carpal tunnel syndrome and other muscular and skeletal disorders are reported. Disinfectant chemicals and infectious bacteria are causes of respiratory illnesses, allergic reactions, diarrhea, and skin infections.

Respiratory Consequences

Poultry housing has been shown to have adverse effects on the respiratory health of workers, ranging from a cough to chronic bronchitis. Workers are exposed to concentrated airborne particulate matter (PM) and endotoxins (a harmful waste product of bacteria. In a conventional hen house a conveyor belt beneath the cages removes the manure. In a cage-free aviary system the manure coats the ground, resulting in the build-up of dust and bacteria over time. Eggs are often laid on the ground or under cages in the aviary housing, causing workers to come close to the floor and force dust and bacteria into the air, which they then inhale during egg collection.

Excretory Consequences

Oxfam America reports that huge industrialized poultry operations are under such pressure to maximize profits that workers are denied access to restrooms.

World Chicken Population

The Food and Agriculture Organization of the United Nations estimated that in 2002 there were nearly sixteen billion chickens in the world, counting a total population of 15,853,900,000. The figures from the *Global Livestock Production and Health Atlas* for 2004 were as follows:

1. China (3,860,000,000)

2. United States (1,970,000,000)

3. Indonesia (1,200,000,000)

4. Brazil (1,100,000,000)

5. Pakistan (691,948,000)

6. India (648,830,000)

7. Mexico (540,000,000)

8. Russia (340,000,000)

9. Japan (286,000,000)

10. Iran (280,000,000)

11. Turkey (250,000,000)

12. Bangladesh (172,630,000)

13. Nigeria (143,500,000)

In 2009 the annual number of chicken raised was estimated at 50 billion, with 6 billion raised in the European Union, over 9 billion raised in the United States and more than 7 billion in China.

In 1950, the average America consumed 20 pounds of chicken per year, but it is predicted that the average consumption will be 89 pounds in 2015. Additionally, in 1980 most chickens were sold whole, and by 2000 almost 90 percent of chickens were sold after being processed into parts. This increase in consumption and processing has led to many of these occupation-related illness.

Poultry

Poultry are domesticated birds kept by humans for the eggs they produce, their meat, their feathers, or sometimes as pets. These birds are most typically members of the superorder Galloanserae (fowl), especially the order Galliformes (which includes chickens, quails and turkeys) and the family Anatidae, in order Anseriformes, commonly known as "waterfowl" and including domestic ducks and domestic geese. Poultry also includes other birds that are killed for their meat, such as the young of pigeons (known as squabs) but does not include similar wild birds hunted for sport or food and known as game.

Poultry of the World

The domestication of poultry took place several thousand years ago. This may have originally been as a result of people hatching and rearing young birds from eggs collected from the wild, but later involved keeping the birds permanently in captivity. Domesticated chickens may have been used for cockfighting at first and quail kept for their songs, but soon it was realised how useful it was having a captive-bred source of food. Selective breeding for fast growth, egg-laying ability, conformation, plumage and docility took place over the centuries, and modern breeds often look very different from their wild ancestors. Although some birds are still kept in small flocks in extensive systems, most birds available in the market today are reared in intensive commercial enterprises. Poultry is the second most widely eaten type of meat globally and, along with eggs, provides nutritionally beneficial food containing high-quality protein accompanied by a low proportion of fat. All poultry meat should be properly handled and sufficiently cooked in order to reduce the risk of food poisoning.

Definition

The word "poultry" comes from the Middle English "pultrie", from Old French *pouletrie*, from *pouletier*, poultry dealer, from *poulet*, pullet. The word "pullet" itself comes from Middle English *pulet*, from Old French *polet*, both from Latin *pullus*, a young fowl, young animal or chicken. The word "fowl" is of Germanic origin (cf. Old English *Fugol*, German *Vogel*, Danish *Fugl*).

"Poultry" is a term used for any kind of domesticated bird, captive-raised for its utility, and traditionally the word has been used to refer to wildfowl (Galliformes) and waterfowl (Anseriformes). "Poultry" can be defined as domestic fowls, including chickens, turkeys, geese and ducks, raised for the production of meat or eggs and the word is also used for the flesh of these birds used as food. The Encyclopædia Britannica lists the same bird groups but also includes guinea fowl and squabs (young pigeons). In R. D. Crawford's *Poultry breeding and genetics*, squabs are omitted but Japanese quail and common pheasant are added to the list, the latter frequently being bred in captivity and released into the wild. In his 1848 classic book on poultry, *Ornamental and Domestic Poultry: Their History, and Management*, Edmund Dixon included chapters on the peafowl, guinea fowl, mute swan, turkey, various types of geese, the muscovy duck, other ducks and all types of chickens including bantams. In colloquial speech, the term "fowl" is often used near-synonymously with "domesticated chicken" (*Gallus gallus*), or with "poultry" or even just "bird", and many languages do not distinguish between "poultry" and "fowl". Both words are also used for the flesh of these birds. Poultry can be distinguished from "game", defined as wild birds or mammals hunted for food or sport, a word also used to describe the flesh of these when eaten.

Examples

Bird	Wild ancestor	Domestication	Utilization	Picture
Chicken	Red junglefowl	Southeast Asia	meat, feathers, eggs, ornamentation, leather	
Duck	Muscovy duck/ Mallard	various	meat, feathers, eggs	
Emu	Emu	various, 20th century	meat, leather, oil	
Egyptian goose	Egyptian goose	Egypt	meat, feathers, eggs, ornamentation	
Goose	Greylag goose/Swan goose	various	meat, feathers, eggs	
Indian peafowl	Indian Peafowl	various	meat, feathers, ornamentation, landscaping	

Mute swan	Mute swan	various	feathers, eggs, landscaping	
Ostrich	Ostrich	various, 20th century	meat, eggs, feathers, leather	
Pigeon	Rock dove	Middle East	meat, feathers, ornamentation	
Quail	Japanese quail, Northern bobwhite	Japan, Virginia	meat, eggs, feathers, pets	
Turkey	Wild turkey	Mexico	meat, feathers	
Grey francolin	Grey francolin	Pakistan, North India	meat, fighting, pets	
Guineafowl	Helmeted guineafowl	Africa	meat, pest consumption, and alarm calling	
Common pheasant	Common pheasant	Eurasia	meat	

Golden pheasant	Golden pheasant	Eurasia	meat, mainly ornamental	
Greater rhea	Greater Rhea	various, 20th century	meat, leather, oil, eggs	

Chickens

Cock with comb and wattles

Chickens are medium-sized, chunky birds with an upright stance and characterised by fleshy red combs and wattles on their heads. Males, known as cocks, are usually larger, more boldly coloured, and have more exaggerated plumage than females (hens). Chickens are gregarious, omnivorous, ground-dwelling birds that in their natural surroundings search among the leaf litter for seeds, invertebrates, and other small animals. They seldom fly except as a result of perceived danger, preferring to run into the undergrowth if approached. Today's domestic chicken (*Gallus gallus domesticus*) is mainly descended from the wild red junglefowl of Asia, with some additional input from grey junglefowl. Domestication is believed to have taken place between 7,000 and 10,000 years ago, and what are thought to be fossilized chicken bones have been found in northeastern China dated to around 5,400 BC. Archaeologists believe domestication was originally for the purpose of cockfighting, the male bird being a doughty fighter. By 4,000 years ago, chickens seem to have reached the Indus Valley and 250 years later, they arrived in Egypt. They were still used for fighting and were regarded as symbols of fertility. The Romans used them in divination, and the Egyptians made a breakthrough when they learned the difficult technique of artificial incubation. Since then, the keeping of chickens has spread around the world for the production of food with the domestic fowl being a valuable source of both eggs and meat.

Since their domestication, a large number of breeds of chickens have been established, but with the exception of the white Leghorn, most commercial birds are of hybrid origin. In about 1800, chickens began to be kept on a larger scale, and modern high-output poultry farms were present in the United Kingdom from around 1920 and became established in the United States soon after the Second World War. By the mid-20th century, the poultry meat-producing industry was of greater importance than the egg-laying industry. Poultry breeding has produced breeds and strains to fulfil different needs; light-framed, egg-laying birds that can produce 300 eggs a year; fast-growing, fleshy birds destined for consumption at a young age, and utility birds which produce both an acceptable number of eggs and a well-fleshed carcase. Male birds are unwanted in the egg-laying industry and can often be identified as soon as they are hatch for subsequent culling. In meat breeds, these birds are sometimes castrated (often chemically) to prevent aggression. The resulting bird, called a capon, has more tender and flavorful meat, as well.

Roman mosaic depicting a cockfight

A bantam is a small variety of domestic chicken, either a miniature version of a member of a standard breed, or a "true bantam" with no larger counterpart. The name derives from the town of Bantam in Java where European sailors bought the local small chickens for their shipboard supplies. Bantams may be a quarter to a third of the size of standard birds and lay similarly small eggs. They are kept by small-holders and hobbyists for egg production, use as broody hens, ornamental purposes, and showing.

Cockfighting

Cockfighting is said to be the world's oldest spectator sport and may have originated in Persia 6,000 years ago. Two mature males (cocks or roosters) are set to fight each other, and will do so with great vigour until one is critically injured or killed. Breeds such as the Aseel were developed in the Indian subcontinent for their aggressive behaviour. The sport formed part of the culture of the ancient Indians, Chinese, Greeks, and Romans, and large sums were won or lost depending on the outcome of an encounter. Cockfighting has been banned in many countries during the last century on the grounds of cruelty to animals.

Ducks

Ducks are medium-sized aquatic birds with broad bills, eyes on the side of the head, fairly long necks, short legs set far back on the body, and webbed feet. Males, known as drakes, are often larger than females (simply known as ducks) and are differently coloured in some breeds. Domestic ducks are omnivores, eating a variety of animal and plant materials such as aquatic insects, molluscs, worms, small amphibians, waterweeds, and grasses. They feed in shallow water by dabbling, with their heads underwater and their tails upended. Most domestic ducks are too heavy to fly, and they are social birds, preferring to live and move around together in groups. They keep their plumage waterproof by preening, a process that spreads the secretions of the preen gland over their feathers.

Pekin ducks

Clay models of ducks found in China dating back to 4000 BC may indicate the domestication of ducks took place there during the Yangshao culture. Even if this is not the case, domestication of the duck took place in the Far East at least 1500 years earlier than in the West. Lucius Columella, writing in the first century BC, advised those who sought to rear ducks to collect wildfowl eggs and put them under a broody hen, because when raised in this way, the ducks "lay aside their wild nature and without hesitation breed when shut up in the bird pen". Despite this, ducks did not appear in agricultural texts in Western Europe until about 810 AD, when they began to be mentioned alongside geese, chickens, and peafowl as being used for rental payments made by tenants to landowners.

It is widely agreed that the mallard (*Anas platyrhynchos*) is the ancestor of all breeds of domestic duck (with the exception of the Muscovy duck (*Cairina moschata*), which is not closely related to other ducks). Ducks are farmed mainly for their meat, eggs, and down. As is the case with chickens, various breeds have been developed, selected for egg-laying ability, fast growth, and a well-covered carcase. The most common commercial breed in the United Kingdom and the United States is the Pekin duck, which can lay 200 eggs a year and can reach a weight of 3.5 kg (7.7 lb) in 44 days. In the Western world, ducks are not as popular as chickens, because the latter produce larger quantities of white, lean meat and are easier to keep intensively, making the price of chicken meat lower than that of duck meat. While popular in *haute cuisine*, duck appears less frequently in the mass-market food industry. However, things are different in the East. Ducks are more popular there than chickens and are mostly still herded in the traditional way and selected for their ability to find sufficient food in harvested rice fields and other wet environments.

Geese

An Emden goose, a descendent of the wild greylag goose

The greylag goose (*Anser anser*) was domesticated by the Egyptians at least 3000 years ago, and a different wild species, the swan goose (*Anser cygnoides*), domesticated in Siberia about a thousand years later, is known as a Chinese goose. The two hybridise with each other and the large knob at the base of the beak, a noticeable feature of the Chinese goose, is present to a varying extent in these hybrids. The hybrids are fertile and have resulted in several of the modern breeds. Despite their early domestication, geese have never gained the commercial importance of chickens and ducks.

Domestic geese are much larger than their wild counterparts and tend to have thick necks, an upright posture, and large bodies with broad rear ends. The greylag-derived birds are large and fleshy and used for meat, while the Chinese geese have smaller frames and are mainly used for egg production. The fine down of both is valued for use in pillows and padded garments. They forage on grass and weeds, supplementing this with small invertebrates, and one of the attractions of rearing geese is their ability to grow and thrive on a grass-based system. They are very gregarious and have good memories and can be allowed to roam widely in the knowledge that they will return home by dusk. The Chinese goose is more aggressive and noisy than other geese and can be used as a guard animal to warn of intruders. The flesh of meat geese is dark-coloured and high in protein, but they deposit fat subcutaneously, although this fat contains mostly monounsaturated fatty acids. The birds are killed either around 10 or about 24 weeks. Between these ages, problems with dressing the carcase occur because of the presence of developing pin feathers.

In some countries, geese and ducks are force-fed to produce livers with an exceptionally high fat content for the production of *foie gras*. Over 75% of world production of this product occurs in France, with lesser industries in Hungary and Bulgaria and a growing production in China. *Foie gras* is considered a luxury in many parts of the world, but the process of feeding the birds in this way is banned in many countries on animal welfare grounds.

Turkeys

Turkeys are large birds, their nearest relatives being the pheasant and the guineafowl. Males are larger than females and have spreading, fan-shaped tails and distinctive, fleshy wattles, called a snood, that hang from the top of the beak and are used in courtship display. Wild turkeys can fly, but seldom do so, preferring to run with a long, stratling gait. They roost in trees and forage on the ground, feeding on seeds, nuts, berries, grass, foliage, invertebrates, lizards, and small snakes.

Male domesticated turkey sexually displaying by showing the snood hanging over the beak, the caruncles hanging from the throat, and the 'beard' of small, black, stiff feathers on the chest

The modern domesticated turkey is descended from one of six subspecies of wild turkey (*Meleagris gallopavo*) found in the present Mexican states of Jalisco, Guerrero and Veracruz. Pre-Aztec tribes in south-central Mexico first domesticated the bird around 800 BC, and Pueblo Indians inhabiting the Colorado Plateau in the United States did likewise around 200 BC. They used the feathers for robes, blankets, and ceremonial purposes. More than 1,000 years later, they became an important food source. The first Europeans to encounter the bird misidentified it as a guineafowl, a bird known as a "turkey fowl" at that time because it had been introduced into Europe via Turkey.

Commercial turkeys are usually reared indoors under controlled conditions. These are often large buildings, purpose-built to provide ventilation and low light intensities (this reduces the birds' activity and thereby increases the rate of weight gain). The lights can be switched on for 24-hrs/day, or a range of step-wise light regimens to encourage the birds to feed often and therefore grow rapidly. Females achieve slaughter weight at about 15 weeks of age and males at about 19. Mature commercial birds may be twice as heavy as their wild counterparts. Many different breeds have been developed, but the majority of commercial birds are white, as this improves the appearance of the dressed carcass, the pin feathers being less visible. Turkeys were at one time mainly consumed on special occasions such as Christmas (10 million birds in the United Kingdom) or Thanksgiving (60 million birds in the United States). However, they are increasingly becoming part of the everyday diet in many parts of the world.

Quail

The quail is a small to medium-sized, cryptically coloured bird. In its natural environment, it is found in bushy places, in rough grassland, among agricultural crops, and in other places with dense cover. It feeds on seeds, insects, and other small invertebrates. Being a largely ground-dwelling, gregarious bird, domestication of the quail was not difficult, although many of its wild instincts are retained in captivity. It was known to the Egyptians long before the arrival of chickens and was depicted in hieroglyphs from 2575 BC. It migrated across Egypt in vast flocks and the birds could sometimes be picked up off the ground by hand. These were the common quail (*Coturnix coturnix*), but modern domesticated flocks are mostly of Japanese quail (*Coturnix japonica*) which was probably domesticated as early as the 11th century AD in Japan. They were originally kept as songbirds, and they are thought to have been regularly used in song contests.

Japanese quail

In the early 20th century, Japanese breeders began to selectively breed for increased egg production. By 1940, the quail egg industry was flourishing, but the events of World War II led to the complete loss of quail lines bred for their song type, as well as almost all of those bred for egg production. After the war, the few surviving domesticated quail were used to rebuild the industry, and all current commercial and laboratory lines are considered to have originated from this population. Modern birds can lay upward of 300 eggs a year and countries such as Japan, India, China, Italy, Russia, and the United States have established commercial Japanese quail farming industries. Japanese quail are also used in biomedical research in fields such as genetics, embryology, nutrition, physiology, pathology, and toxicity studies. These quail are closely related to the common quail, and many young hybrid birds are released into the wild each year to replenish dwindling wild populations.

Other Poultry

Guinea fowl originated in southern Africa, and the species most often kept as poultry is the helmeted guineafowl (*Numida meleagris*). It is a medium-sized grey or speckled bird with a small naked head with colourful wattles and a knob on top, and was domesticated by the time of the ancient Greeks and Romans. Guinea fowl are hardy, sociable birds that subsist mainly on insects, but also consume grasses and seeds. They will keep a vegetable garden clear of pests and will eat the ticks that carry Lyme disease. They happily roost in trees and give a loud vocal warning of the approach of predators. Their flesh and eggs can be eaten in the same way as chickens, young birds being ready for the table at the age of about four months.

A squab is the name given to the young of domestic pigeons that are destined for the table. Like other domesticated pigeons, birds used for this purpose are descended from the rock pigeon (*Columba livia*). Special utility breeds with desirable characteristics are used. Two eggs are laid and incubated for about 17 days. When they hatch, the squabs are fed by both parents on "pigeon's milk", a thick secretion high in protein produced by the crop. Squabs grow rapidly, but are slow to fledge and are ready to leave the nest at 26 to 30 days weighing about 500 g (18 oz). By this time, the adult pigeons will have laid and be incubating another pair of eggs and a prolific pair should produce two squabs every four weeks during a breeding season lasting several months.

Poultry Farming

Free-range ducks in Hainan Province, China

Worldwide, more chickens are kept than any other type of poultry, with over 50 billion birds being raised each year as a source of meat and eggs. Traditionally, such birds would have been kept extensively in small flocks, foraging during the day and housed at night. This is still the case in developing countries, where the women often make important contributions to family livelihoods through keeping poultry. However, rising world populations and urbanization have led to the bulk of production being in larger, more intensive specialist units. These are often situated close to where the feed is grown or near to where the meat is needed, and result in cheap, safe food being made available for urban communities. Profitability of production depends very much on the price of feed, which has been rising. High feed costs could limit further development of poultry production.

In free-range husbandry, the birds can roam freely outdoors for at least part of the day. Often, this is in large enclosures, but the birds have access to natural conditions and can exhibit their normal behaviours. A more intensive system is yarding, in which the birds have access to a fenced yard and poultry house at a higher stocking rate. Poultry can also be kept in a barn system, with no access to the open air, but with the ability to move around freely inside the building. The most intensive system for egg-laying chickens is battery cages, often set in multiple tiers. In these, several birds share a small cage which restricts their ability to move around and behave in a normal manner. The eggs are laid on the floor of the cage and roll into troughs outside for ease of collection. Battery cages for hens have been illegal in the EU since January 1, 2012.

Chickens raised intensively for their meat are known as "broilers". Breeds have been developed that can grow to an acceptable carcass size (2 kg (4.4 lb)) in six weeks or less. Broilers grow so fast, their legs cannot always support their weight and their hearts and respiratory systems may not be able to supply enough oxygen to their developing muscles. Mortality rates at 1% are much higher than for less-intensively reared laying birds which take 18 weeks to reach similar weights. Processing the birds is done automatically with conveyor-belt efficiency. They are hung by their feet, stunned, killed, bled, scalded, plucked, have their heads and feet removed, eviscerated, washed, chilled, drained, weighed, and packed, all within the course of little over two hours.

Both intensive and free-range farming have animal welfare concerns. In intensive systems, cannibalism, feather pecking and vent pecking can be common, with some farmers using beak trimming as a preventative measure. Diseases can also be common and spread rapidly through the flock. In extensive systems, the birds are exposed to adverse weather conditions and are vulnerable to predators and disease-carrying wild birds. Barn systems have been found to have the worst bird welfare. In Southeast Asia, a lack of disease control in free-range farming has been associated with outbreaks of avian influenza.

Poultry Shows

In many countries, national and regional poultry shows are held where enthusiasts exhibit their birds which are judged on certain phenotypical breed traits as specified by their respective breed standards. The idea of poultry exhibition may have originated after cockfighting was made illegal, as a way of maintaining a competitive element in poultry husbandry. Breed standards were drawn up for egg-laying, meat-type, and purely ornamental birds, aiming for uniformity. Sometimes, poultry shows are part of general livestock shows, and sometimes they are separate events such as the annual "National Championship Show" in the United Kingdom organised by the Poultry Club of Great Britain.

Poultry as Food

Trade

Chicken and duck eggs on sale in Hong Kong

Poultry is the second most widely eaten type of meat in the world, accounting for about 30% of total meat production worldwide compared to pork at 38%. Sixteen billion birds are raised annually for consumption, more than half of these in industrialised, factory-like production units. Global broiler meat production rose to 84.6 million tonnes in 2013. The largest producers were the United States (20%), China (16.6%), Brazil (15.1%) and the European Union (11.3%). There are two distinct models of production; the European Union supply chain model seeks to supply products which can be traced back to the farm of origin. This model faces the increasing costs of implementing additional food safety requirements, welfare issues and environmental regulations. In contrast, the United States model turns the product into a commodity.

World production of duck meat was about 4.2 million tonnes in 2011 with China producing two thirds of the total, some 1.7 billion birds. Other notable duck-producing countries in the Far East include Vietnam, Thailand, Malaysia, Myanmar, Indonesia and South Korea (12% in total). France (3.5%) is the largest producer in the West, followed by other EU nations (3%) and North America (1.7%). China was also by far the largest producer of goose and guinea fowl meat, with a 94% share of the 2.6 million tonne global market.

Global egg production was expected to reach 65.5 million tonnes in 2013, surpassing all previous years. Between 2000 and 2010, egg production was growing globally at around 2% per year, but since then growth has slowed down to nearer 1%.

Cuts of Poultry

In the poultry pavilion of the Rungis International Market, France

Poultry is available fresh or frozen, as whole birds or as joints (cuts), bone-in or deboned, seasoned in various ways, raw or ready cooked. The meatiest parts of a bird are the flight muscles on its chest, called "breast" meat, and the walking muscles on the legs, called the "thigh" and "drumstick". The wings are also eaten (Buffalo wings are a popular example in the United States) and may be split into three segments, the meatier "drumette", the "wingette" (also called the "flat"), and the wing tip (also called the "flapper"). In Japan, the wing is frequently separated, and these parts are referred to as 手羽元 (*teba-moto* "wing base") and 手羽先 (*teba-saki* "wing tip").

Dark meat, which avian myologists refer to as "red muscle", is used for sustained activity—chiefly walking, in the case of a chicken. The dark colour comes from the protein myoglobin, which plays a key role in oxygen uptake and storage within cells. White muscle, in contrast, is suitable only for short bursts of activity such as, for chickens, flying. Thus, the chicken's leg and thigh meat are dark, while its breast meat (which makes up the primary flight muscles) is white. Other birds with breast muscle more suitable for sustained flight, such as ducks and geese, have red muscle (and therefore dark meat) throughout. Some cuts of meat including poultry expose the microscopic regular structure of intracellular muscle fibrils which can diffract light and produce iridescent colours, an optical phenomenon sometimes called structural colouration.

Health and Disease (Humans)

Cuts from plucked chickens

Poultry meat and eggs provide nutritionally beneficial food containing protein of high quality. This is accompanied by low levels of fat which have a favourable mix of fatty acids. Chicken meat contains about two to three times as much polyunsaturated fat as most types of red meat when measured by weight. However, for boneless, skinless chicken breast, the amount is much lower. A 100-g serving of baked chicken breast contains 4 g of fat and 31 g of protein, compared to 10 g of fat and 27 g of protein for the same portion of broiled, lean skirt steak.

A 2011 study by the Translational Genomics Research Institute showed that 47% of the meat and poultry sold in United States grocery stores was contaminated with *Staphylococcus aureus*, and 52% of the bacteria concerned showed resistance to at least three groups of antibiotics. Thorough cooking of the product would kill these bacteria, but a risk of cross-contamination from improper handling of the raw product is still present. Also, some risk is present for consumers of poultry meat and eggs to bacterial infections such as *Salmonella* and *Campylobacter*. Poultry products may become contaminated by these bacteria during handling, processing, marketing, or storage, resulting in food-borne illness if the product is improperly cooked or handled.

In general, avian influenza is a disease of birds caused by bird-specific influenza A virus that is not normally transferred to people; however, people in contact with live poultry are at the greatest risk of becoming infected with the virus and this is of particular concern in areas such as Southeast Asia, where the disease is endemic in the wild bird population and domestic poultry can become infected. The virus possibly could mutate to become highly virulent and infectious in humans and cause an influenza pandemic.

Bacteria can be grown in the laboratory on nutrient culture media, but viruses need living cells in which to replicate. Many vaccines to infectious diseases can be grown in fertilised chicken eggs. Millions of eggs are used each year to generate the annual flu vaccine requirements, a complex process that takes about six months after the decision is made as to what strains of virus to include in the new vaccine. A problem with using eggs for this purpose is that people with egg allergies are unable to be immunised, but this disadvantage may be overcome as new techniques for cell-based rather than egg-based culture become available. Cell-based culture will also be useful in a pandemic when it may be difficult to acquire a sufficiently large quantity of suitable sterile, fertile eggs.

Free Range

Free range denotes a method of farming husbandry where the animals, for at least part of the day, can roam freely outdoors, rather than being confined in an enclosure for 24 hours each day. On many farms, the outdoors ranging area is fenced, thereby technically making this an enclosure, however, free range systems usually offer the opportunity for extensive locomotion and sunlight prevented by indoor housing systems. *Free range* may apply to meat, eggs or dairy farming.

The term is used in two senses that do not overlap completely: as a farmer-centric description of husbandry methods, and as a consumer-centric description of them. There is a diet where the practitioner only eats meat from free-range sources called ethical omnivorism, which is a type of semivegetarian.

In ranching, free-range livestock are permitted to roam without being fenced in, as opposed to fenced-in pastures. In many of the agriculture-based economies, free-range livestock are quite common.

History

If one allows "free range" to include "herding", free range was a typical husbandry method at least until the development of barbed wire and chicken wire. The generally poor understanding of nutrition and diseases before the twentieth century made it difficult to raise many livestock species without giving them access to a varied diet, and the labor of keeping livestock in confinement and carrying all their feed to them was prohibitive except for high-profit animals such as dairy cattle.

In the case of poultry, free range was the dominant system until the discovery of vitamins A and D in the 1920s, which allowed confinement to be practised successfully on a commercial scale. Before that, green feed and sunshine (for the vitamin D) were necessary to provide the necessary vitamin content. Some large commercial breeding flocks were reared on pasture into the 1950s. Nutritional science resulted in the increased use of confinement for other livestock species in much the same way.

United States

In the United States, USDA free range regulations currently apply only to poultry and indicate that the animal has been allowed access to the outside. The USDA regulations do not specify the quality or size of the outside range nor the duration of time an animal must have access to the outside.

The term "free range" is mainly used as a marketing term rather than a husbandry term, meaning something on the order of, "low stocking density," "pasture-raised," "grass-fed," "old-fashioned," "humanely raised," etc.

There have been proposals to regulate by the USDA the labeling of products as free range within the United States. As of now what constitutes raising an animal free range is entirely decided by the producer of that product

Free-range Poultry

Free range meat chickens seek shade on a U.S. farm

In poultry-keeping, "free range" is widely confused with yarding, which means keeping poultry in fenced yards. Yarding, as well as floorless portable chicken pens ("chicken tractors") may have some of the benefits of free-range livestock but, in reality, the methods have little in common with the free-range method.

A behavioral definition of free range is perhaps the most useful: "chickens kept with a fence that restricts their movements very little." This has practical implications. For example, according to Jull, "The most effective measure of preventing cannibalism seems to be to give the birds good grass range." De-beaking was invented to prevent cannibalism for birds not on free range, and the need for de-beaking can be seen as a litmus test for whether the chickens' environment is sufficiently "free-range-like."

The U.S. Department of Agriculture Food Safety and Inspection Service (FSIS) requires that chickens raised for their meat have access to the outside in order to receive the free-range certification. There is no requirement for access to pasture, and there may be access to only dirt or gravel . Free-range chicken eggs, however, have no legal definition in the United States. Likewise, free-range egg producers have no common standard on what the term means.

The broadness of "free range" in the U.S. has caused some people to look for alternative terms. "Pastured poultry" is a term promoted by farmer/author Joel Salatin for broiler chickens raised on grass pasture for all of their lives except for the initial brooding period. The Pastured Poultry concept is promoted by the American Pastured Poultry Producers' Association (APPPA), an organization of farmers raising their poultry using Salatin's principles.

Free-range Livestock

Traditional American usage equates "free range" with "unfenced," and with the implication that there was no herdsman keeping them together or managing them in any way. Legally, a free-range jurisdiction allowed livestock (perhaps only of a few named species) to run free, and the owner was not liable for any damage they caused. In such jurisdictions, people who wished to avoid damage by livestock had to fence them out; in others, the owners had to fence them in.

The USDA has no specific definition for "free-range" beef, pork, and other non-poultry products. All USDA definitions of "free-range" refer specifically to poultry.

In a December 2002 Federal Register notice and request for comments (67 Fed. Reg. 79552), USDA's Agricultural Marketing Service proposed "minimum requirements for livestock and meat industry production/marketing claims". Many industry claim categories are included in the notice, including breed claims, antibiotic claims, and grain fed claims. "Free Range, Free Roaming, or Pasture Raised" would be defined as "livestock that have had continuous and unconfined access to pasture throughout their life cycle" with an exception for swine ("continuous access to pasture for at least 80% of their production cycle"). In a May 2006 Federal Register notice (71 Fed. Reg. 27662), the agency presented a summary and its responses to comments received in the 2002 notice, but only for the category "grass (forage) fed" which the agency stated was to be a category separate from "free range." Comments received for other categories, including "free range," are to be published in future Federal Register editions.

European Union

A free range chicken flock in the USA

The European Union regulates marketing standards for egg farming which specifies the following (cumulative) minimum conditions for the free-range method:

- hens have continuous daytime access to open-air runs, except in the case of temporary restrictions imposed by veterinary authorities,

- the open-air runs to which hens have access is mainly covered with vegetation and not used for other purposes except for orchards, woodland and livestock grazing if the latter is authorised by the competent authorities,

- the open-air runs must at least satisfy the conditions specified in Article 4(1)(3)(b)(ii) of Directive 1999/74/EC whereby the maximum stocking density is not greater than 2500 hens per hectare of ground available to the hens or one hen per $4m^2$ at all times and the runs are not extending beyond a radius of 150 m from the nearest pophole of the building; an extension of up to 350 m from the nearest pophole of the building is permissible provided that a sufficient number of shelters and drinking troughs within the meaning of that provision are evenly distributed throughout the whole open-air run with at least four shelters per hectare.

Free range geese in Germany

Otherwise, egg farming in EU is classified into 4 categories: Organic (ecological), Free Range, Barn, and Cages.) The mandatory labelling on the egg shells attributes a number (which is the first digit on the label) to each of these categories: 0 for Organic, 1 for Free Range, 2 for Barn and 3 for Cages.

There are EU regulations about what free-range means for laying hens and broilers (meat chickens) as indicated above. However, there are no EU regulations for free-range pork, so pigs could be indoors for some of their lives. In order to be classified as free-range, animals must have access to the outdoors for at least part of their lives.

United Kingdom

Free range pigs in England

Pigs Free-range pregnant sows are kept in groups and are often provided with straw for bedding, rooting and chewing. Around 40% of UK sows are kept free-range outdoors and farrow in huts on their range.

Egg laying hens Cage-free egg production includes barn, free-range and organic systems. In the UK, free-range systems are the most popular of the non-cage alternatives, accounting for around 28% of all eggs, compared to 4% in barns and 6% organic. In free-range systems, hens are housed to a similar standard as the barn or aviary.

Free-range rearing of pullets Free range rearing of pullets for egg-laying is now being pioneered in the UK by various poultry rearing farms. In these systems, the pullets are allowed outside from as young as 4 weeks of age, rather than the conventional systems where the pullets are reared in barns and allowed out at 16 weeks of age

Meat chickens Free-range broilers are reared for meat and are allowed access to an outdoor range for at least 8 hours each day. Free-range broiler systems use slower-growing breeds of chicken to improve welfare, meaning they reach slaughter weight at 16 weeks of age rather than 5–6 weeks of age in standard rearing systems.

Turkeys Free-range turkeys have continuous access to an outdoor range during the daytime. The range should be largely covered in vegetation and allow more space. Access to fresh air and daylight means better eye and respiratory health. The turkeys are able to exercise and exhibit natural behaviour resulting in stronger, healthier legs. Free-range systems often use slower-growing breeds of turkey.

Australia

Australian standards in relation to free-range production are largely espoused in third-party certification trade marks due to the absence of any significant legally binding legislation. A number of certification bodies are utilised by rearers to identify their products with a particular level of animal welfare standards. In events where producers do not choose to use a certified trade mark and merely state that their product is 'free range', the producer is bound by consumer expectations and perceptions of what constitutes free range. Producers are generally thought to be bound to Model Codes of Practice of Animal Welfare published by the CSIRO, and in some states this forms part of legislation.

Egg Laying Hens

In Australia, three farming methods for the production of eggs are utilised. In 2011, traditional cage (or battery) eggs accounted for 42% of value, barn-laid eggs account for 10% of value, free-range eggs accounted for 44% of value, and organic eggs accounted for 4% of value. Increased demand for free range eggs due to customer concerns over animal welfare has led to a number of different standards developing in relation to three core welfare measures - indoor stocking density, outdoor stocking density, and beak trimming. The Model Code of Practice recommends practices for free range farming with the following standards:

- Maximum stocking densities indoors of 30 kg/m2, equivalent to about 14-15 birds/m2.

- Maximum outdoor stocking density of 1500 birds/ha, although this can be increased with rotation onto fresh pasture

- Access to the outdoor range for a minimum of 8 hours per day, except in adverse weather conditions

- 2 metres worth of popholes per 1,000 birds for access to the range

- Beak trimming is permitted, and to be undertaken by an accredited operator

The above standards are not always met, and on some occasions producers may want more ethical standards. As such, certified trade marks play a significant role in the determining of what constitutes free range. The key certifications used for layer hens in Australia include the following...

Egg Corp Assured is the weakest standard, set by the industry peak group and largely based on the Model Code of Practice. Egg Corp Assured differs in that it interprets the outdoor stocking density figure as largely irrelevant to welfare. Egg Corp Assured has been known to certify farms running up to 44,000 birds per hectare outdoors, far in excess of recommendations. Like the Model Code of practice, beak trimming is allowed and indoor densities run up to 15 birds per m2.

RSPCA Approved Farming is a standard that can be applied to both barn-laid and free-range egg producers. Farms using this certification must have an indoor density of 9 birds/m2 indoors on slats, or 7 birds/m2 indoors in a deep-litter system. The standards dictate a maximum outdoor density of 1500 per ha without rotation, or 2500 birds per ha with rotation, and beak trimming is allowed.

Free Range Egg & Poultry Australia (FREPA) standards provides a sliding scale for indoor density, with 10 birds/m2 allowed only in enclosures housing less than 1,000 birds, and 6 birds/m2 the maximum for barns with over 4,000 birds. Nothing is said in the standards about outdoor density, thus it is assumed that farmers must meet the standards under the Model Code. Beak trimming is allowed under this certification.

Humane Choice True Free Range standards are some of the most sound as far as animal welfare is concerned. Beak trimming or any other mutilations is not permitted, perches must be provided, and maximum flock numbers cannot be greater than 2,500 per barn. The outdoor stocking density is 1500 birds per ha, and the indoor density is 5 birds per m2.

Australian Certified Organic Standards include criteria on feed content and the use of pesticides in addition to animal welfare requirements. The indoor density is a maximum of 8 birds/m2, although most operators under this standard list their density as 5 birds per m2. The outdoor density is 1000 birds per ha, and beak trimming is not permitted.

Chicken Meat

In Australia, free range and organic chicken accounts for about 16.6% of value in the poultry market. This percentage is expected to grow to up to 25% in the next 5 years. No meat birds are raised in cages in Australia. There are three main certification trademarks in this market.

Free Range Egg & Poultry Australia (FREPA) standards are those in which most supermarket brands of free range chicken meat are accredited under. These standards require indoor stocking densities of up to 30 kg per m2 indoors (about 15 birds per m2), and beak trimming is not permitted. Outdoor stocking density is not stated, but it is understood that the outdoor range must be at a minimum 1.5 times the floor area of inside the barn.

RSPCA Approved Farming standards for free range require an indoor stocking density of about 17 birds per m2, and outdoor densities of up to 17 birds per m2. No beak trimming is allowed under this system.

Australian Certified Organic standards dictate a maximum indoor stocking density of up to 12 birds per m2 indoors, and 2500 birds per hectare outdoors. These standards require perches, and prevent large, conventional broiler sheds.

Yarding

During the daytime, the doors are left open for these chickens to choose whether to be in the yard or coop. This small poultry farm is in Hainan, China.

In poultry keeping, yarding is the practice of providing the poultry with a fenced yard in addition to a poultry house. Movable yarding is a form of managed intensive grazing.

Yarding is often confused with free range. The distinction is that free-range poultry are either totally unfenced, or the fence is so distant that it has little influence on their freedom of movement.

Historical Practice

Before the discovery of vitamins A and D in the 1920s, green feed and sunshine were essential to the health of poultry. Vitamin D was synthesized from sunlight on the skin (as with humans), while Vitamin A was obtained through green forage plants such as grass. Yards small enough to be fenced economically were soon stripped of palatable green forage and become barren. This is followed by a build-up of manure, parasites, and other pathogens.

Free range husbandry was the most common method in these early days. Most farms had only a small free-range barnyard flock. Larger flocks were kept in small houses build on skids, which were dragged periodically to a fresh piece of ground. This method is similar to the modern practice of pastured poultry.

Experts of the day estimated the sustainable level to be about fifty hens per acre (80 m² per hen), with one hundred hens per acre (40 m² per hen) as an absolute upper limit if special care was taken. These levels are sustainable in the sense that the turf can make use of the nutrients in the manure left behind by the chickens, and in the sense that, at this stocking density, the chickens will not completely destroy the turf through scratching.

At the Oregon Station on clay soil it was found that the day droppings from 200 laying hens on an acre [20 m² per hen] in four years made the soil too rich for the successful growth of cereal crops where cropping the ground was done every other year. The night droppings were put on other land. If the soil contains too much manure for the crops it is safe to assume that it is not in the best condition for poultry. Sooner or later it is bound to show not only in a failure of grain crops but in failure of poultry crops. For a permanent system under average conditions of soil and climate the following points are suggested for consideration.

- Maximum number of fowls per acre: 100 laying hens [40 m² per hen].

- Disposing of the night droppings on other land.

- Dividing the ground into at least two divisions or yards, and growing a crop on each yard at least every other year. In sections where crops may be grown every year the number of fowls may be increased.

- Growing crops that will use up the maximum amount of manure.

- Keeping the ground vacant [of chickens] at least six months in the year.

- Thorough underdrainage, where necessary, to carry off surplus water.

.... It is not assumed that as many as 500 hens may not be profitably kept on an acre [8 m² per hen] for a few years under favorable conditions. It has been done, but it is a different matter when it is planned to make a permanent business of it.

Because fifty hens per acre represents 800 square feet (74 m²) per hen (80 m² per hen), while the density inside the house at the time was normally four square feet per hen (0.4 m² per hen), this required that the yard be 200 times wider than the house, assuming a yard on one side of the house. That is, a house 20 feet (6 m) wide required a yard 4,000 feet (1,220 m) wide to provide the necessary area. This would normally be provided as two yards, one on either side of the house, each 2,000 feet (610 m) wide. In reality, such yards are expensive to fence, and the chickens spend most of their time on the portion closest to the house, so sustainability was never achieved in practice except with portable houses, which were moved periodically to fresh ground. Yarded operations were operated with unsustainably small yards that were quickly denuded and which received excessive levels of manure.

The use of multiple yards, frequent plowing, and liberal use of lime would allow higher stocking levels to be used, since plowing and liming would allow much of the nitrogen to escape from the soil.

The following is typical advice for the successful use of yards in the Thirties and Forties:

All poultrymen should realize that there are no known substitutes for sunshine and young green grass in keeping poultry in the best possible state of health and in promoting growth and maintaining egg production. Where sunshine and green grass cannot be provided, as in the case of birds kept in strict confinement, the best possible substitutes must be provided. In the case of most farm and many commercial flocks, however, the growing stock is reared on range, and the adult birds are given yards or allowed to roam at will.

If the staggering losses among growing chicks and laying birds that occur annually are to be reduced materially, better methods of flock management must be employed. The losses from mortality are due largely to internal parasites and diseases of one kind or another. Bare ground over which the chickens have run for some time, mud puddles, and stagnant water are the chief sources of the spread of diseases, most of which are filth borne....

The mortality that usually occurs in growing and adult stock may be materially reduced by providing the birds with an alternate yarding system. Probably the best arrangement is to provide each colony brooder and each laying house with three yards (3 m) which the birds would be allowed to use every 3 or 4 weeks. By alternating the birds in the yards every 3 or 4 weeks each yard is kept reasonably sanitary, especially if the soil in the immediate vicinity of the house is cultivated and treated with lime, and young green grass is available for the birds throughout the season. The importance of clean range for both birds and adult stock cannot be emphasized too strongly... For adult stock a good grass sward can be maintained on fertile soil, allowing about 200 birds to the acre [20 m^2 per hen].

Nutritional advances increasingly turned yarding into a liability, and it fell out of favor. Free range continued to be used, especially for breeding flocks and for pullets before they reached laying age, because of the lower rate of disease and greater overall health of grass-reared chickens. Breeding flocks (which lay eggs destined for incubation) are always given a better diet than flocks laying table eggs, since a diet that will produce table eggs cheaply will not provide eggs that hatch well. For some time after confined laying flocks produced table eggs satisfactorily, breeding flocks benefited from free range

In Britain, Geoffrey Sykes developed a new yarding system in the Fifties. This used a small yard covered with a thick layer of straw, with more straw added frequently. He also recommended that shade and a windbreak be provided by a solid fence around the yard, or by other means, such as rows of haybales. Once a year, the old straw was removed by a front-end loader or similar machinery. This method eliminated mud and pathogens. It was later forgotten because the industry moved to high-density confinement before the method was widely established.

Recent Practice

Today, commercial poultry producers generally call yarding free range on their labels. This conflation of two very different techniques has led to confusion. The vast majority of "free-range" operations are really yarded.

Pastured poultry, as promoted by the APPPA, the American Pastured Poultry Producers Association, and author/farmer Joel Salatin, takes a different approach, attempting to achieve the benefits of free range while using penning or yarding. The key element of Pastured poultry is the use of portable housing and the optional use of portable electric fencing. By moving the house and yard frequently, perhaps daily, all the disadvantages of permanent yards are eliminated.

Battery Cage

Chickens in battery cages showing individual cages.

Battery cages are a housing system used for various animal production methods, but primarily for egg-laying hens. The name arises from the arrangement of rows and columns of identical cages connected together, sharing common divider walls, as in the cells of a battery. Although the term is usually applied to poultry farming, similar cage systems are used for other animals. Battery cages have generated controversy between advocates for animal rights, and industrial producers.

It was estimated that over 60% of the world's eggs were produced in industrial systems, mostly using battery cages. In the US, over 90% of the 300 million egg-laying chickens are housed in battery cages. In the UK, statistics from the Department for the Environment, Food and Rural Affairs (Defra) indicate that 50% of eggs produced in the UK throughout 2010 were from cages (45% from free-range, 5% from barns). However, introduction of the European Union Council Directive 1999/74/EC which banned conventional battery cages in the EU from January 2012 for welfare reasons, means the number of eggs from battery cages in the EU states is rapidly decreasing. The EU ban was proposed when international scientists independently observed signs of extreme abnormal behaviour (including cannibalism) in caged hens.

Examples

Battery cages also used for mink, rabbit, chinchilla and foxes in fur farming, and most recently for the Asian palm civet for kopi luwak production of coffee.

Battery cages for sun bears reared for their bile

Battery cages for mink reared for their fur

Battery cages for silver foxes reared for their fur

Battery cages for civets reared for kopi luwak (coffee) production

History

Before the cage was invented, most hens were free-range.

A chicken coop from the 1950s

An early reference to battery cages appears in Milton Arndt's 1931 book, *Battery Brooding*, where he reports that his cage flock was healthier and had higher egg production than his conventional flock. At this early date, battery cages already had the sloped floor that allowed eggs to roll to the front of the cage, where they were easily collected by the farmer and out of the hens' reach. Arndt also mentions the use of conveyor belts under the cages to remove manure, which provides better air control quality and eliminates fly breeding.

Original battery cages extended the technology used in battery brooders, which were cages with a wire mesh floor and integral heating elements for brooding chicks. The wire floor allowed the manure to pass through, removing it from the chicks' environment and reducing the risk of manure-borne diseases.

Early battery cages were often used for selecting hens based on performance, since it is easy to track how many eggs each hen is laying if only one hen is placed in a cage. Later, this was combined with artificial insemination, giving a technique where each egg's parentage is known. This method is still used today.

Early reports from Arndt about battery cages were enthusiastic. Arndt reported:

"This form of battery is coming into widespread use throughout the country and apparently is solving a number of the troubles encountered with laying hens in the regular laying house on the floor.

In the first edition of this book I spoke of my experimental work with 220 pullets which were retained for one year in individual cages. At the end of this year it was found that the birds confined in the batteries outlaid considerably the same size flock in the regular houses. The birds consume less feed than those on the floor and this coupled with the increased production made them more profitable than the same number of pullets in the laying house.

A number of progressive poultrymen from all over the United States and some in foreign countries cooperated with me in carrying on experimental work with this type of battery and each and every one of them were very well satisfied with the results obtained. In fact, a number of them have since placed their entire laying flocks in individual hen batteries."

In 1967, Samuel Duff filed a patent for "battery cages" in patent US3465722.

The use of laying batteries increased gradually, becoming the dominant method somewhat before the integration of the egg industry in the 1960s. The practice of battery cages was criticized in Ruth Harrison's landmark book *Animal Machines*, published in 1964.

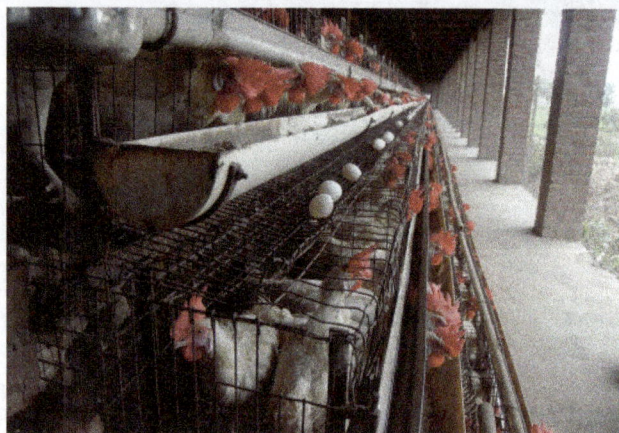

A simple battery cage system with no conveyors for feed or eggs

In 1990, North and Bell reported that 75% of all commercial layers in the world and 95% in the United States were kept in cages.

By all accounts, a caged layer facility is more expensive to build than high-density floor confinement, but can be cheaper to operate if designed to minimize labor.

North and Bell report the following economic advantages to laying cages:

1. It is easier to care for the pullets; no birds are underfoot. 2. Floor eggs are eliminated. 3. Eggs are cleaner. 4. Culling is expedited. 5. In most instances, less feed is required to produce a dozen eggs. 6. Broodiness is eliminated. 7. More pullets may be housed in a given house floor space. 8. Internal parasites are eliminated. 9. Labor requirements are generally much reduced

They also cite disadvantages to cages:

1. The handling of manure may be a problem. 2. Generally, flies become a greater nuisance. 3. The investment per pullet may be higher than in the case of floor operations. 4. There is a slightly higher percentage of blood spots in the eggs. 5. The bones are more fragile and processors often discount the fowl price.

Disadvantages 1 and 2 can be eliminated by manure conveyors, but some industrial systems do not feature manure conveyors.

Legislation

European Union

In 1999, the European Union Council Directive 1999/74/EC banned the conventional battery cage in the EU from 2012, after a 12-year phase-out. In their 1996 report, the European Commission's Scientific Veterinary Committee (SVC) condemned the battery cage, concluding:

"It is clear that because of its small size and its barrenness, the battery cage as used at present has inherent severe disadvantages for the welfare of hens".

The EU Directive allows enriched or "furnished" cages to be used. Under the directive, enriched cages must be at least 45 cm high and must provide each hen with at least 750 cm² of space; 600 cm² of this must be "usable area" – the other 150 cm² is for a nest-box. The cage must also contain litter, perches and "claw-shortening devices". Some animal welfare organisations, such as Compassion in World Farming, have criticised this move, calling for enriched cages to be prohibited as they believe they provide no significant or worthwhile welfare benefits compared with conventional battery cages.

Germany banned conventional battery cages from 2007, five years earlier than required by the EU Directive, and has prohibited enriched cages from 2012. Mahi Klosterhalfen of the Albert Schweitzer Foundation has been instrumental in a strategic campaign against battery cages in Germany.

Switzerland

Switzerland banned battery cages from January 1, 1992; it was the first country to impose such a ban.

United States

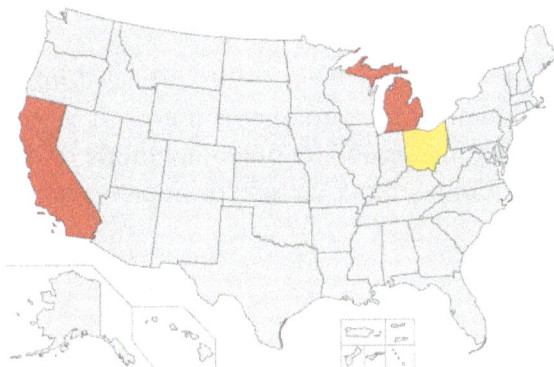

US States with bans on battery cages for laying hens

| Laws prohibiting battery cages

| Moratorium on permits for new battery cage construction

The passage of California Proposition 2 (2008) aimed, in part, to reduce or eliminate the problems associated with battery cages, by setting the standard for space relative to free movement and wingspan, rather than cage size.

Battery cages are also illegal in Michigan due to HB 5127, passed in 2009, which mandates that certain farm animals have enough room to stand up, lie down, turn around and extend their limbs, rather than being confined in tiny cages.

Finally, in Ohio, there is a moratorium on permits for the construction of new battery cages as of June 2010.

Australia

The 2009 'Code of Practice' permits the use of battery cages. A written commitment by the Federal government to review the practice was scheduled in 2010; there was no further communication. During 2013 the state government of Tasmania was planning to phase out battery cages and budgeting for financial compensation for affected farmers but this was scrapped following the 2014 election. Attempts to change the law have been an object of contention; RSPCA Australia officially campaigns against cage eggs.

Welfare Concerns

There are several welfare concerns regarding the battery cage system of housing and husbandry. These are presented below in the approximate chronological order they would influence the hens.

Chick Culling

Due to modern selective breeding, laying hen strains are different from those of meat production strains. Male birds of the laying strains do not lay eggs and are unsuitable for meat production, therefore, they are culled soon after being sexed, often on the day of hatching. Methods of culling include cervical dislocation, asphyxiation by carbon dioxide and maceration using a high speed grinder.

Animal rights groups have used videos of live chicks being placed into macerators as evidence of cruelty in the egg production industry. Maceration, together with cervical dislocation and asphyxiation by carbon dioxide, are all considered acceptable methods of euthanasia by the American Veterinary Medical Association. Consumers may also be appalled simply by the death of animals that are not subsequently eaten.

Beak-trimming

To reduce the harmful effects of feather pecking, cannibalism and vent pecking, most chicks eventually going into battery cages are beak-trimmed. This is often performed on the first day after hatching, simultaneously with sexing and receiving vaccinations. Beak-trimming is a procedure considered by many scientists to cause acute pain and distress with possible chronic pain; it is practised on chicks for all types of housing systems, not only battery cages.

Cage Size

At approximately 16 weeks of age, pullets (hens which have not yet started to lay) are placed into cages. In countries with relevant legislation, floor space for battery cages ranges upwards from 300 cm² per bird. EU standards in 2003 called for at least 550 cm² per hen. In the US, the current recommendation by the United Egg Producers is 67 to 86 in² (430 to 560 cm²) per bird. The space available to each hen in a battery cage has often been described as less than the size of a sheet of A4 paper (624 cm²). Others have commented that a typical cage is about the size of a filing cabinet drawer and holds eight to 10 hens.

Battery cage

Behavioural studies showed that when turning, hens used 540 to 1006 cm², when stretching wings 653 to 1118 cm², when wing flapping 860 to 1980 cm², when feather ruffling 676 to 1604 cm², when preening 814 to 1240 cm², and when ground scratching 540 to 1005 cm². A space allowance of 550 cm² would prevent hens in battery cages from performing these behaviours without touching another hen. Animal welfare scientists have been critical of battery cages because of these space restrictions and it is widely considered that hens suffer boredom and frustration when unable to perform these behaviours. Spatial restriction can lead to a wide range of abnormal behaviours, some of which are injurious to the hens or their cagemates.

Light Manipulation

To reduce the harmful effects of feather pecking, cannibalism and vent-pecking, hens in battery cages (and other housing systems) are often kept at low light intensities (e.g. less than 10 lux). Low light intensites may be associated with welfare costs to the hens as they prefer to eat in brightly lit environments and prefer brightly lit areas for active behaviour but dim (less than 10 lux) for in-active behaviour. Dimming the lights can also cause problems when the intensity is then abruptly increased temporarily to inspect the hens; this has been associated as a risk factor of increased feather pecking and the birds can become frightened resulting in panic-type ("hysteria") reactions which can increase the risk of injury.

Being indoors, hens in battery cages do not see sunlight. Whilst there is no scientific evidence for this being a welfare problem, some animal advocates indicate it is a concern. Furnished cages and some other non-cage indoor systems would also prevent hens seeing natural light throughout their lives.

Osteoporosis

Several studies have indicated that toward the end of the laying phase (approximately 72 weeks of age), a combination of high calcium demand for egg production and a lack of exercise can lead to osteoporosis. This can occur in all housing systems for egg laying hens, but is particularly preva-lent in battery cage systems where it has sometimes been called 'cage layer osteoporosis'. Osteopo-rosis leads to the skeleton becoming fragile and an increased risk of bone breakage, particularly in the legs and keel bone. Fractures may occur whilst the hens are in the cage and these are usually discovered at depopulation as old, healed breaks, or they might be fresh breaks which occurred during the process of depopulation. One study showed that 24.6% of hens from battery cages had recent keel fractures whereas hens in furnished cages, barn and free-range had 3.6%, 1.2% and 1.3% respectively. However, hens from battery cages experienced fewer old breaks (17.7%) compared to hens in barn (69.1%), free-range (59.8%) and furnished cages (31.7%).

Forced Moulting

Flocks are sometimes force moulted, rather than being slaughtered, to reinvigorate egg-laying. This involves complete withdrawal of food (and sometimes water) for 7 to 14 days or sufficiently long to cause a body weight loss of 25 to 35%. This stimulates the hen to lose her feathers, but also reinvigorates egg-production. Some flocks may be force moulted several times. In 2003, more than 75% of all flocks were moulted in the US. This temporary starving of the hens is seen as inhumane and is the main point of objection by critics and opponents of the practice. The alternative most often employed is to slaughter the hens instead of moulting them.

Improving Welfare for Egg Producing Hens

The Scientific Veterinary Committee of the European Commission stated that "enriched cages and well designed non-cage systems have already been shown to have a number of welfare advantages over battery systems in their present form". Supporters of battery husbandry contend that alternative systems such as free range also have welfare problems, such as increases in cannibalism, feather pecking and vent pecking. A recent review of welfare in battery cages made the point that such welfare issues are problems of management, unlike the issues of behavioural deprivation, which are inherent in a system that keeps hens in such cramped and barren conditions. Free range egg producers can limit or eliminate injurious pecking, particularly feather pecking, through such strategies as providing environmental enrichment, feeding mash instead of pellets, keeping roosters in with the hens, and arranging nest boxes so hens are not exposed to each other's vents; similar strategies are more restricted or impossible in battery cages.

Furnished Cages

Furnished cages, sometimes called "enriched" or "modified" cages, are cages for egg laying hens which have been designed to overcome some of the welfare concerns of battery cages whilst retaining their economic and husbandry advantages, and also provide some of the welfare advantages of non-cage systems. Many design features of furnished cages have been incorporated because research in animal welfare science has shown them to be of benefit to the hens.

History and Legislation

Battery cages are already banned in several countries including Switzerland, Belgium, Austria, Sweden and the Netherlands and prototype commercial furnished cage systems were being developed in the 1980s. In 1999, the European Union Council Directive 1999/74/EC banned the conventional battery cage in the EU from 2012, after a 10-year phase-out. As alternatives to battery cages, the EU Council Directive allowed non-cage systems and furnished cages. Furnished cages therefore represent a feasible alternative to battery cages in the EU after 2012.

Under Directive 1999/74/EC, furnished cages must provide at least the following: 750 cm^2 per hen, of which 600 cm^2 is 45 cm high, a nest, a littered area for scratching and pecking, 15 cm of perch and 12 cm of food trough per hen and a claw shortening device.

Austria banned battery cages in 2009 and is set to ban furnished cages by 2020. Belgium has also banned the battery cage – and proposes to ban furnished cages by 2024. Germany has introduced a 'family cage', which has more space than the furnished cages used in other countries, however, consumers in Germany have been rejecting these eggs. Outside the EU, Switzerland has already banned both the battery and furnished cage systems.

All major UK supermarkets have promised to stop selling eggs from furnished cages by 2025.

Furnished Cages and Battery Cages

Furnished cages retain several advantages of battery cages in that they-

- Separate the eggs from the hens' feces thereby keeping the eggs clean

- Protect the hens from predation

- Automatically collect the eggs thereby preventing egg-eating and floor-laying which both incur additional cost

- Retain a small group size which reduces injurious pecking behaviour

Furnished cages have welfare benefits additional to battery cages by providing -

- Additional space

- A nest

- A claw shortening device

- A dust bath/litter substrate

- A perch

- Easier access for depopulation

Current Designs

There is no clear limit to the size of the furnished cages. Although initial models were not much larger than conventional battery cages, most current designs house 40 to 80 hens although one system houses 115 hens. The depth of furnished cages is often more than the depth of battery cages and as a result, they are often arranged with only one cage row per level, i.e. not connected back-to-back. The more shallow cages can be connected back-to-back. To create space for large groups of hens, some designs of furnished cages are very long. Cage bottoms are made of wire mesh or plastic slats and are sloped so that eggs not laid in the nest box roll onto an egg belt. Feed is provided in feeders outside the cage, although in some designs there may be internal feeders or a combination of the two. Perches in some designs are raised and in others are at floor level.

Welfare Benefits

In a study which compared the welfare benefits of hens in furnished cages, battery cages, free range and barn systems, hens in furnished cages had the lowest faecal corticosterone (a hormone that indicates stress levels), the lowest number of hens that were vent pecked, lowest number of egg shells with calcium spots (an indicator of stress when the egg is temporarily retained by the hen), lowest number of egg shells with blood spots on (usually caused by prolapse), lowest score of skin damage, lowest severity of vent damage caused by vent pecking and lowest plumage soiling. Hens in furnished cages had a similar percentage of hens with recent keel fractures which are usually caused during depopulation (3.6%) compared to hens in barn (1.2%) and free-range systems (1.3%), all of which were considerably lower than in hens from battery cages (24.6%). Furthermore, hens in furnished cages had a smaller percentage of old keel fractures (31.7%) compared to hens in barn (69.1%) and free-range (59.8%) systems but more than hens in battery cages (17.7%). This indicates that furnished cages protect against the keel breaks that are common amongst non-caged hens and also protects against the effects of osteoporosis prevalent in battery cages causing bones to be weak and easily broken during depopulation. In this study, mortality rates were above the breed standards in all systems except the furnished cages.

Welfare Disadvantages

Furnished cages provide more space than battery cages but still prevent some behaviours such as vigorous wing-flapping, flying, nest-building (no materials are provided) and inhibit others (comfort or grooming behaviours) determined partly by the numbers of hens in the cage. The hens are not separated from their feces as completely as hens in battery cages and therefore are at a greater risk of disease, although not as great as the risk to hens in non-cage systems. The small amount of litter that is provided in furnished cages is often distributed quickly or flicked out the cage, possibly resulting in frustration for hens wishing to dustbath and resulting in sham dustbathing. The nest boxes are often occupied by hens using the box for behaviours other than egg-laying (e.g. for sleeping or sham dustbathing) which could lead to frustration in hens wishing to lay an egg.

Production in Furnished Cages

Some studies indicate that production in furnished cages is comparable to that in battery cages. Other studies indicate hens housed in furnished cages have better bodyweights and egg production compared to hens in battery cages.

Broiler Industry

The Broiler industry is the process by which broiler chickens are reared and prepared for meat consumption.

A broiler chicken in Ecuador

Broiler Industry Structure

The broiler production process is very much an industrial one. There are several distinct components of the broiler supply chain.

Primary Breeding Sector

The "primary breeding sector" consists of companies that breed pedigree stock. Pedigree stock ("pure line") is kept on high level biosecure farms. Their eggs are hatched in a special pedigree hatchery and their progeny then goes on to the great grandparent (GGP) and grandparent (GP) generations. These eggs would then go to a special GP hatchery to produce Parent Stock (PS) which passes to the production sector.

In 2006, out of an estimated world population of 18 billion poultry, about 3% are breeding stock. The US supplied about 1/4 of world GP stock.

Worldwide, the primary sector produced 417 million parent stock (PS) per year.

A single pedigree-level hen might have 25000 parent stock bird descendents, which in turn might produce 3 million broilers.

Numerous techniques are used to assess the pedigree stock. For example, birds might be examined with ultrasound or x-rays to study the shape of muscles and bones. The blood oxygen level is measured to determine cardiovascular health. The walking ability of pedigree candidates is observed and scored.

The need for high levels of R&D spending prompted consolidation within the primary breeder industry. By the late 2000s only three sizable breeding groups remained:

- Aviagen (with the Ross, Arbor Acres, Indian River and Peterson brands)

- Cobb-Vantress (with the Cobb, Avian, Sasso and Hybro brands), and

- Groupe Grimaud (with the Hubbard and Grimaud Frere brands).

In the UK, 2 international firms supply about 90% of the parent stock.

Due to the high levels of variation in the chicken genome, the industry has not yet reached biological limits to improved performance.

The full chicken genome was published in Nature, in December 2004. Today, all primary breeding groups are investing heavily in genomics research. This research mostly focuses on understanding the function and effect of genes already present in the breeding population. Research into transgenics — removing genes or artificially moving genes from one individual or species to another — has fewer prospects of gaining favor among consumers.

Broiler Breeder (Parent Stock) Farms

Broiler breeder farms raise parent stock which produce fertilized eggs. A broiler hatching egg is never sold at stores and is not meant for human consumption. The males and females are separate genetic lines or breeds. The chicks they produce will therefore be hybrids or 'crosses'. Since the birds are bred mainly for efficient meat production, producing eggs can be a challenge. In Canada, the average producer houses 15,000 birds that begin laying hatching eggs at 26 weeks of age. Each bird will lay about 150 hatching eggs for the next 34 to 36 weeks. This cycle is then repeated when the producer puts another flock of 26 week-old birds into his barns to begin the process again. As a general rule, each farmer produces enough broiler hatching eggs to supply chicks for 8 chicken producers. (Other sources indicate a parent hen will lay about 180 eggs in 40 week production period.)

Generally, parent flocks are either owned by integrated broiler companies or hatcheries or are contracted to them on a long-term basis.

Broiler breeder growing is typically a two-stage process. Parent stock purchased from a primary breeder is delivered as day old. Most are first placed with on specialist *rearing houses* or starter farms until approximately 18 weeks of age. The starter farm has the specialized brooding equipment to raise the chicks.

Rearing House

A typical rearing house (also called a shed or barn) design for Alabama-like climate (100 °F (38 °C) in summer and 20 °F (−7 °C) in winter):

- 40 by 400 feet (12 m × 122 m) size, single storey.

- 11,000 bird capacity (about 1.4 sq ft (0.13 m²) per bird)

- Ceiling is insulated

- Exterior curtain side walls.

- A "minimum ventilation" system is required for the heating period to provide a certain amount of fresh air.

- A separate "tunnel ventilation" system with evaporative pad cooling is desired (minimum wind speed is 400 fpm) for hot weather in the later stage of the bird's growth.

- Air inlets may be automatically adjusted.

- A negative ventilation system helps keep dirt and dust out of egg storage areas.

- The entire house may be heated, or individual "brooders" may be used.

- The floor is flat. There are no "slats" or "pits" for manure. There are no cages, and no nests. "Litter" (shavings or straw) covers the floor. When the chicks are introduced temporary barriers are used to keep them close to the heated areas.

- "Black-out" design to keep out external light, so the day-night cycle can be controlled.

- An automatic timer-controlled lighting system. Dimmers allow light intensity to be adjusted.

- Automatic feeders to distribute feed. Typically this consists of an endless chain in a trough or with individual pans. A silo or bin outside provides storage.

- Automatic drinkers provide water. There are several different designs, with "nipples" or "round" drinkers being popular.

- Feeders and drinkers are height adjusted as the birds grow, and can be raised on chains or wires to allow cleanout of the barn.

Chicks require warm air temperatures, which is reduced as the birds mature:

Age	Brooder Temperature	Whole-House Heating Temperature
0 days	34–35 °C (93-95 °F)	31–32 °C (88-90 °F)
14 days	31–32 °C (88-90 °F)	24–25 °C (75-77 °F)

Chicks might be debeaked at 7–10 days age. During rearing, bird weight is carefully monitored, as an over-weight bird will be a poor egg producer. The feed mix will be adjusted to meet nutritional needs at each stage. Feed might be restricted to control body weight, for example with "skip a day" feeding, or feeding 5 days out of 7. A vaccination program is carried out, which ensures the longevity of the parent stock, and the immunity may be passed to the broiler progeny. Males (cockerels) and females (pullets), are usually raised separately.

Laying House

The birds are then moved to broiler breeder *laying houses* or production barns. The birds are typically placed into crates, and transported by truck to a separate facility. Males and females are raised together at this point. Outwardly the laying house will resemble the rearing house. Inside, about one-half of the floor might consist of raised 'slats.' During the production run, manure will drop through the slats and accumulate in the pit underneath the slats. The birds are not generally caged, especially since the roosters must mate with the hens to fertilize the eggs. Nests are provided for laying hens. Both automatic and manual (example) nesting systems exist. Manual nests are usually stuffed with straw or shavings and eggs are hand-collected. Automatic systems usually have a plastic carpet lining, with a belt for egg collection. Careful layout and attention to bird behavior is required to avoid 'floor eggs'.

Depending on breed, egg production starts at 24–26 weeks of age. Production percentage (daily eggs per hen) climbs rapidly to a peak of 80–85% at 29–32 weeks, and then gradually declines with age. Hatchability tends to peak (at perhaps 90%) somewhat later than production at 34–36 weeks. Overall flock production will decline as mortality reduces the size of the flock.

When the rooster mates with the hen, sperm enter the hen's oviduct and are stored within sperm storage glands. These glands can store more than half a million sperm, and sperm can remain viable for up to 3 weeks. However, a hen will have maximum fertility for only about 3 to 4 days after one mating. Therefore, the male-to-female ratio in a flock must be enough to ensure mating of every hen every 3 days or so. To maintain fertility, younger roosters may be introduced as the flock ages.

Eggs are collected a minimum of twice a day, and usually more frequently. Cracked or dirty eggs are separated, as they are not suitable for hatching. Undersized, oversized or double-yolk eggs are also unsuitable. The eggs might be disinfected by fumigation, are packed in 'flats' or trays, placed in wheeled trolleys, and stored in a cool (15-18C) climate-controlled area. The egg packing room and storage rooms are kept segrated to reduce contamination. The trolleys are delivered by truck to a hatchery perhaps twice a week.

At the end of the production cycle, the birds are called "spent fowl". Disposal of spent fowl may be a problem as consumer demand for them is poor.

Hatcheries

Five-day-old broiler strain Cornish-Rock chicks

Hatcheries take the fertilized eggs, incubate them, and produce day old broiler chicks.

Incubation takes about 21 days, and is often a two-step process. Initial incubation is done in machines known as *setters*. A modern setter is the size of a large room, with a central corridor and racks on either side. Eggs are held relatively tightly (large end up) in trays, which are stored in the racks. Inside the setter, temperature and humidity are closely maintained. Blowers or fans circulate air to ensure uniform temperature, and heating or cooling is applied as needed by automated controls. The racks pivot or tilt from side to side, usually on an hourly basis. As an example, one commercial machine can hold up to 124,416 eggs and measures about 4.7 metres wide by 7.2 metres deep. Setters often hold more than one hatch, on a staggered hatch-day basis, and operate continuously. The setter phase lasts about 18 days.

On or about day 18, the eggs are removed from the setters and transferred to *hatchers*. These machines are similar to setters, but have larger flat-bottom trays, so the eggs can rest on their sides, and newly hatched chicks can walk. Having a separate machine helps keep hatching debris out of the setter. The environmental conditions in the hatcher are optimized to help the chicks hatch. As a commercial example, a large hatcher has capacity for 15,840 eggs, and measures about 3.3 metres by 1.8 metres.

Some incubators are single-stage (combining setter and hatcher funcations), and entire trolleys of eggs can be rolled in at one time. One advantage of single-stage machines is that they are thoroughly cleaned after each hatch, whereas a setter is rarely shutdown for cleaning. The single-stage environment can be adjusted for eggs from different producing flocks, and for each step of the hatch cycle. The setter environment is often a compromise as different egg batches are in the machine at one time.

On hatch day (day 21), the trays are removed ("pulled") from the hatchers, and then the chicks are removed from the trays. Chicks are inspected, with the sickly ones being disposed of. Chicks may be by vaccinated, sorted by sex, counted, and placed in chick boxes.(Example1) (Example2) Stacks of chick boxes are loaded into trucks for transport, and arrive at the broiler farm on the same day. Specialized climate-controlled trucks are typically used, depending on climate and transport distance.

Chick sexing is optionally done to improve uniformity – since males grow faster, the weights are more uniform if males and females are raised separately. The birds are bred so that males and females have unique feather patterns or color differences. Unlike egg-laying poultry, males are not culled.

Typical hatchability rate in Canada in 2011 was 82.2%. (i.e. 82.2% of eggs set for incubation produced a saleable chick). A UK source estimates 90% hatchability.

Broiler Farms

Broiler chickens in a farm

The chicks are delivered to the actual broiler *Grow-Out* farms. In the US, houses may be up to 60' x 600' (36000 sq.ft.). One 2006 magazine survey reported a desired 67 foot wide house, with the average 'standard' new house being 45' x 493', with largest being 60' x 504'. One farm complex may have several houses.

In Mississippi, typical farms now have four to six houses with 25,000 birds per house. One full-time worker might manage three houses. On average, a new broiler house is about 500 feet long by 44 feet wide and costs about $200,000 equipped.

When the birds are full-grown, they are caught (perhaps with a Chicken harvester) placed in crates, and transported by truck to a processing plant.

Broiler chickens kept outside near a chicken shop in India

Because of their efficient meat conversion, broiler chickens are also popular in small family farms in rural communities, where a family will raise a small flock of broilers.

Processing Plants

When the birds are large enough, they are shipped to processing plants for slaughter. When chickens arrive at the processor they go through the following sequence:

- Removed from transport cages

- Hung by the legs on a shackle, mounted on a conveyor chain.

- Stunned using an electrically charged water bath

- Killed by cutting the blood vessels in the neck

- Bled so that most blood has left the carcass

- Scalded to soften the attachment of the feathers

- Plucked to remove the feathers

- Head removed

- Hock cutting to remove the feet

- Rehung in the evisceration room

- Gutted or eviscerated to remove the internal organs

- Washed to remove blood and soiling from the carcass

- Chilled to prevent bacterial spoiling (They go through a chiller which takes approximately 2 hours to go through. The chiller generally holds thousands of gallons of water kept below 40 degrees Fahrenheit.)

- Drained to allow excess water to drip off the carcass

- Weighing

- Cut selection to divide the carcass into desired portion (breast, drumsticks etc.)

- Packed (for example in plastic bags) to protect carcasses or cuts

- Chilled or frozen for preservation

Further Processing plants carry out operations such as cutting and deboning. Previously the conveyor belts carrying live chickens generally ran at a maximum of 140 chickens per minute, but the maximum speed has been increased to 175 birds/minute. Once the dead birds arrive in the evisceration room (usually dropped down a chute after the feet are removed), they are hung again on shackles much the same way as they were when they were alive.

Feed Mills

Integrators

Today, in the U.S. an individual company called an "integrator" performs all or most production aspects. Integrators generally own breeder flocks, hatcheries, feed mills, and processing plants. The integrators provide the chicks, feed, medication, part of the fuel for brooding, and technical advisers to supervise farm production. Integration reduces costs by coordinating each stage of production.

U.S. Industry History

In the 1920s–1930s, broiler production was initiated in locations such as the Delmarva Peninsula, Georgia, Arkansas, and New England. Mrs. Wilmer Steele of Sussex County, Delaware, is often cited as the pioneer of the commercial broiler industry. In 1923, she raised a flock of 500 chicks intended to be sold for meat. Her business was so profitable that by 1926 she was able to build a 10,000 bird broiler house.

In 1945, A&P organized the first of its "Chicken of Tomorrow" contests. Qualifying trials were conducted in 1946 and 1947 with the national finals held in 1948. Breeders submitted a case of 30 dozen hatching eggs to a hatchery, the eggs were hatched, the offspring raised until they reached market weight and were then slaughtered. Broilers were judged on several factors, including growth rate, feed conversion efficiency, and the amount of meat on breasts and drumsticks. Though held only three times, the contests enabled breeders such as Peterson, Vantress, Cobb, Hubbard, Pilch and Arbor Acres to become established market brands.

During the 1940s – 1960s, feed mills, hatcheries, farms, and processors were all separate entities. Hatcheries were driven to co-ordinate activities to protect their market share and production. Later, feed mills extended credit to farmers to purchase feed to produce the live chickens. Eventually entrepreneurs consolidated feed mill, hatchery and processing operations, resulting in the beginnings of the integrated industry.

Chickens were typically sold "New York dressed," with only the blood and feathers removed. In 1942, an Illinois plant was the first to win government approval of "on-line" evisceration. Evisceration and ice-packing of ready-to-cook whole carcasses became the norm. In 1949, USDA launched a voluntary program of grading. Federal inspection of broilers became mandatory in 1959.

By 1952, "broilers" surpassed farm chickens as the number one source of chicken meat in the United States.

By the mid-1960s, ninety percent of broilers produced came from vertical integrated operations, which combined production, processing and marketing.

In the late 1960s and early 1970s, major companies used television and print media to market chickens under brand names. Today, 95 percent of broilers sold at retail grocery stores carry a brand name.

By the early 1980s, consumers preferred cut-up and further-processed chickens to the traditional whole bird.

Chicken passed pork consumption in 1985. Chicken consumption surpassed beef consumption in 1992.

	Historic	**Modern**
Number of Hatcheries	11,405 (1934)	323 (2001)
Incubator Capacity	276 million eggs (1934)	862 million eggs (2001)
Hatchery Average Incubation Capacity	24,224 eggs (1934)	2.7 million eggs (2001)
Annual Broiler Production	366 million broilers (1945)	8.4 billion broilers (2001)
Average live weight	3.03 pounds (1945)	5.06 pounds (2001)
Live weight price	36 cents per pound (1948)	39.3 cents per pound (2001)
Feed Conversion Efficiency	4.70 lbs feed per lb live weight (1925)	1.91 lbs feed per lb live weight (2011)
Mortality	18% (1925)	3.8% (2011)

Breeders

Cobb claims to be world's oldest poultry breeding company. Founded 1916 when Robert C. Cobb Senior purchased a farm in Littleton, Massachusetts, forming Cobb's Pedigreed Chicks. Purchased by Upjohn in 1974. Sold to Tyson Foods in 1994.

Hubbard was founded by Oliver Hubbard in 1921 in Walpole, New Hampshire. Acquired by Merck in 1974. In 1997 Hubbard was spun off and merged with the ISA-group from France as part of Merial. In 2003, split from ISA, while keeping the broiler lines from ISA and Shaver. Sold by Merial to Groupe Grimaud in 2005.

Arbor Acres was originally a family farm, started by Italian immigrant Frank Saglio who purchased a Glastonbury, Connecticut farm in 1917. He started raising chickens in abandoned piano crates. His third son Henry Saglio took over the poultry while in grade eight. Henry began trying to breed a white bird, because black pinfeathers were difficult to pluck. In 1948, and again in 1951, Arbor Acres White Rocks won in the purebred category of the Chicken of Tomorrow competition. The white feathered Arbor Acres birds were preferred to the higher performing dark feathered Red Cornish crosses. In 1964, Nelson Rockefeller purchased Arbor Acres making it part of International Basic Economy Corporation (IBEC). Joint ventures were formed in Thailand, Taiwan, Indonesia, India, the Philippines, and Japan. In 1980, IBEC merged with Booker McConnell Limited of Great Britain. Booker owned all of "AA" by 1991. At this time Arbor Acres had grown to become the world's largest broiler breeding company, with customers in over 70 countries. AA was divested in 2000, eventually acquired by Avigen.

Shaver started with 2 hens in 1932 by Donald Shaver. Mainly focused on laying hens, Shaver launched a broiler product in 1958. Cargill purchased part of Shaver in 1964, which helped give Shaver a toehold in the US market. In the early 1970s the market share in the US was around 8-10%. Cargill bought all of Shaver in 1985. Shaver was acquired by ISA in 1988, and then made part of Merial. The layer business kept the Shaver name, and was sold as Natexis Industrie in 2003, and then to Hendrix in 2005.

Industry Statistics

World

Worldwide, from 1985 to 2005, the broiler industry grew by 158%. Major increases were experienced by:

- China +591%
- Brazil +482%
- US +147%
- Thailand +141%
- EU-25 +73%.

In 2005 world production was 71,851,000 tonnes. Major producers were:

- United States 15,869,000 tonnes
- China 10,196,000 tonnes
- EU-25 8,894,000 tonnes
- Brazil 8,668,000 tonnes

In 2005, world exports of chicken meat $8.3 billion (CAD). Largest exporters were Brazil ($4 billion), the United States ($2.6 billion) and the EU-25 ($0.82 billion). The largest importers of chicken meat were: Japan ($1 billion), Russia ($943.3 million), Germany ($800.6 million) and Hong Kong ($598.8 million).

United States

In 2010, approximately 36.9 billion pounds (16,737,558 tonnes) of broilers were sold, for a retail value of $45 billion, based on retail weight sold multiplied by the retail composite price. In 2010, the US exported 6.8 billion pounds, valued at $3.1 billion, about 18% of production.

In 2009, the US produced 8.6 billion birds. The top 3 states were Georgia, Arkansas and Alabama, each producing over 1 billion birds. Farm receipts were about $22 billion.

There are fewer than 50 highly specialized, vertically integrated agribusiness firms that dominate the industry. The top 10 integrators produce about 60% of U.S. broilers.

In 2001, there were 323 chicken hatcheries, with an incubator capacity of 862 million eggs. The average capacity per hatchery was 2.7 million eggs.

In 2010 the largest producers were Tyson Foods (161 million ready to cook pounds) and Pilgrim's Pride (126.5 million pounds). The next largest producer, Perdue Farms, is less than half the size of Pilgrims Pride.

Canada

Canada has a supply management system where marketing boards govern the broiler and broiler hatching egg industries. For broilers, prices are negotiated at the provincial level. In each province, the minimum price per kg that processors will pay to producers is set periodically through negotiations between processors and the provincial marketing board. From 1992 to 2003, negotiated prices in Ontario are generally used as a benchmark when conducting price negotiations in other provinces. In Ontario, Chicken Farmers of Ontario (CFO) has price-negotiating authority. It negotiates the base price paid by primary processors for live chicken with primary processors. Since 2003, the live chicken price is determined by a "live price formula" established by the Agriculture, Food, and Rural Affairs Appeals Tribunal that includes the price of chicks, feed and producer margin.

Broiler hatching egg production consists of 270 producers generating about $188.3 million in 2005. Canada produced about 675 million hatching eggs, and imported about 121 million.

There were 66 hatcheries in Canada, of which 20 were mixed, producing both broiler and layer chicks. The main companies involved in broiler hatching eggs and chicks are:

- Maple Leaf Foods Incorporated,
- Lilydale Hatchery,
- Maple Lodge,
- Couvoir Boire & Frères Inc
- Western Hatchery Limited.

The average price per chick was about $0.35. Canada imported about 13 million broiler chicks.

There were 2786 regulated chicken producers, generating farm cash receipts of $1.6 billion in 2005. Compared to other livestock sectors (i.e. beef, dairy, and pork), the poultry and egg industry was the healthiest with regards to total income for the average operator.

In 2005, total chicken slaughters were 973.9 million kilograms. Of this, 35.2 million kg were mature (non-broiler)slaughters, meaning about 96% of chicken consumption was broilers. By revenue, chicken processing is about 1/4 of the meat packing business. The top 8 processors account 66% of the market.

In 2005 there were 175 primary poultry processing plants. The five largest firms are, in order:

1. la Coopérative fédérée de Québec (three plants in Québec),

2. Lilydale Poultry Co-operative (one plant in British Columbia, three in Alberta and one in Saskatchewan),

3. Maple Leaf Poultry (two plants in Ontario, one in Alberta and one in Nova Scotia),

4. Exceldor (two plants in Québec) and

5. Maple Lodge Farms (one plant in Ontario).

There are 376 plants that do further processing, involving canning, boning and cutting.

Chicken consumption by market sector	Consumption (000,000 kg)
Retail (Grocery Stores, Butcher Shops	625
Fast Food	231
Full Serve Restaurants	97
Hotels, Institutions	55
Total	1008

The Canadian Food Inspection Agency (CFIA), is a branch of Health Canada whose role is to enforce the food safety, to ensure animal health, to set standards and carry out enforcement and inspection. Activities range from the inspection of federally registered meat processing facilities to border inspections for foreign pests and diseases, to the enforcement of practices related to fraudulent labelling. The CFIA also verifies the humane transportation of animals, conducts food investigations and recalls, and performs laboratory testing and environmental feeds.

The Canadian On-Farm Food Safety Program (COFFSP) is directed by producers. It is a science-based, credible program consistent with the Hazard Analysis Critical Control Points (HACCP, "hass ap") standards, managed by the Canadian Federation of Agriculture.

References

- Lester R. Brown (2003). "Chapter 8. Raising Land Productivity: Raising protein efficiency". Plan B: Rescuing a Planet Under Stress and a Civilization in Trouble. NY: W.W. Norton & Co. ISBN 0-393-05859-X

- Grandin, Temple; Johnson, Catherine (2005). Animals in Translation. New York, NY: Scribner. p. 183. ISBN 0-7432-4769-8.

- Appleby, M.C.; J.A. Mench; B.O. Hughes (2004). Poultry Behaviour and Welfare. Wallingford and Cambridge MA: CABI Publishing. ISBN 0-85199-667-1.

- Cherry, Peter; Morris, T. R. (2008). Domestic Duck Production: Science and Practice. CABI. pp. 1–7. ISBN 978-1-84593-441-5.

- Smith, Andrew F. (2006). The Turkey: An American Story. University of Illinois Press. pp. 4–5, 17. ISBN 978-0-252-03163-2.

- Pond, Wilson, G.; Bell, Alan, W. (eds.) (2010). Turkeys: Behavior, Management and Well-Being. Marcell Dekker. pp. 847–849. ISBN 0-8247-5496-4.

- Hosking, Richard (1996). 日本料理用語辞典 (英文): Ingredients & Culture. Tuttle Publishing. p. 156. ISBN 978-0-8048-2042-4.

- North, Mack O.; Donald E. Bell (1990). Commercial Chicken Production Manual (4th ed.). Van Nostrand Reinhold. pp. 297, 315. ISBN 0-87055-446-8.

- Appleby, M.C.; Mench, J.A.; Hughes, B.O. (2004). Poultry Behaviour and Welfare. Wallingford and Cambridge MA: CABI Publishing. ISBN 0-85199-667-1.

- Brulliard, Karin (10 June 2016). "Egg producers pledge to stop grinding newborn male chickens to death". The Washington Post. Retrieved 12 June 2016.

- Ogle, Maureen. "Riots, Rage, and Resistance: A Brief History of How Antibiotics Arrived on the Farm". Scientific American. Sep 3, 2013. Retrieved 28 October 2016.

5

Organic Farming: An Essential Aspect

Organic farming consciously promotes agricultural practices that do not harm the environment. Organic farming mainly relies on organic fertilizers such as manure, green manure and bone meal. Some of the important aspects dealt within this section are organic horticulture, natural farming, organic certification and organic fertilizer. The topics elaborated in this text will help in gaining a better perspective on organic farming.

Organic Farming

Organic farming is an alternative agricultural system which originated early in the 20th century in reaction to rapidly changing farming practices. Organic agriculture continues to be developed by various organic agriculture organizations today. It relies on fertilizers of organic origin such as compost, manure, green manure, and bone meal and places emphasis on techniques such as crop rotation and companion planting. Biological pest control, mixed cropping and the fostering of insect predators are encouraged. In general, organic standards are designed to allow the use of naturally occurring substances while prohibiting or strictly limiting synthetic substances. For instance, naturally occurring pesticides such as pyrethrin and rotenone are permitted, while synthetic fertilizers and pesticides are generally prohibited. Synthetic substances that are allowed include, for example, copper sulfate, elemental sulfur and Ivermectin. Genetically modified organisms, nanomaterials, human sewage sludge, plant growth regulators, hormones, and antibiotic use in livestock husbandry are prohibited. Reasons for advocation of organic farming include real or perceived advantages in sustainability, openness, self-sufficiency, autonomy/independence, health, food security, and food safety, although the match between perception and reality is continually challenged.

World map of organic agriculture (hectares)

Organic agricultural methods are internationally regulated and legally enforced by many nations, based in large part on the standards set by the International Federation of Organic Agriculture

Movements (IFOAM), an international umbrella organization for organic farming organizations established in 1972. Organic agriculture can be defined as:

Vegetables from ecological farming.

an integrated farming system that strives for sustainability, the enhancement of soil fertility and biological diversity whilst, with rare exceptions, prohibiting synthetic pesticides, antibiotics, synthetic fertilizers, genetically modified organisms, and growth hormones.

Since 1990 the market for organic food and other products has grown rapidly, reaching $63 billion worldwide in 2012. This demand has driven a similar increase in organically managed farmland that grew from 2001 to 2011 at a compounding rate of 8.9% per annum. As of 2011, approximately 37,000,000 hectares (91,000,000 acres) worldwide were farmed organically, representing approximately 0.9 percent of total world farmland.

History

Agriculture was practiced for thousands of years without the use of artificial chemicals. Artificial fertilizers were first created during the mid-19th century. These early fertilizers were cheap, powerful, and easy to transport in bulk. Similar advances occurred in chemical pesticides in the 1940s, leading to the decade being referred to as the 'pesticide era'. These new agricultural techniques, while beneficial in the short term, had serious longer term side effects such as soil compaction, erosion, and declines in overall soil fertility, along with health concerns about toxic chemicals entering the food supply. In the late 1800s and early 1900s, soil biology scientists began to seek ways to remedy these side effects while still maintaining higher production.

Biodynamic agriculture was the first modern system of agriculture to focus exclusively on organic methods. Its development began in 1924 with a series of eight lectures on agriculture given by Rudolf Steiner. These lectures, the first known presentation of what later came to be known as organic agriculture, were held in response to a request by farmers who noticed degraded soil conditions and a deterioration in the health and quality of crops and livestock resulting from the use of chemical fertilizers. The one hundred eleven attendees, less than half of whom were farmers, came from six countries, primarily Germany and Poland. The lectures were published in November 1924; the first English translation appeared in 1928 as *The Agriculture Course*.

In 1921, Albert Howard and his wife Gabrielle Howard, accomplished botanists, founded an Institute of Plant Industry to improve traditional farming methods in India. Among other things, they brought improved implements and improved animal husbandry methods from their scientific training; then by incorporating aspects of the local traditional methods, developed protocalls for the rotation of crops, erosion prevention techniques, and the systematic use of composts and manures. Stimulated by these experiences of traditional farming, when Albert Howard returned to Britain in the early 1930s he began to promulgate a system of natural agriculture.

In July 1939, Ehrenfried Pfeiffer, the author of the standard work on biodynamic agriculture (*Bio-Dynamic Farming and Gardening*), came to the UK at the invitation of Walter James, 4th Baron Northbourne as a presenter at the Betteshanger Summer School and Conference on Biodynamic Farming at Northbourne's farm in Kent. One of the chief purposes of the conference was to bring together the proponents of various approaches to organic agriculture in order that they might cooperate within a larger movement. Howard attended the conference, where he met Pfeiffer. In the following year, Northbourne published his manifesto of organic farming, *Look to the Land*, in which he coined the term "organic farming." The Betteshanger conference has been described as the 'missing link' between biodynamic agriculture and other forms of organic farming.

In 1940 Howard published his *An Agricultural Testament*. In this book he adopted Northbourne's terminology of "organic farming." Howard's work spread widely, and he became known as the "father of organic farming" for his work in applying scientific knowledge and principles to various traditional and natural methods. In the United States J.I. Rodale, who was keenly interested both in Howard's ideas and in biodynamics, founded in the 1940s both a working organic farm for trials and experimentation, The Rodale Institute, and the Rodale Press to teach and advocate organic methods to the wider public. These became important influences on the spread of organic agriculture. Further work was done by Lady Eve Balfour in the United Kingdom, and many others across the world.

Increasing environmental awareness in the general population in modern times has transformed the originally supply-driven organic movement to a demand-driven one. Premium prices and some government subsidies attracted farmers. In the developing world, many producers farm according to traditional methods that are comparable to organic farming, but not certified, and that may not include the latest scientific advancements in organic agriculture. In other cases, farmers in the developing world have converted to modern organic methods for economic reasons.

Terminology

Biodynamic agriculturists, who based their work on Steiner's spiritually-oriented anthroposophy, used the term "organic" to indicate that a farm should be viewed as a living organism, in the sense of the following quotation:

"An organic farm, properly speaking, is not one that uses certain methods and substances and avoids others; it is a farm whose structure is formed in imitation of the structure of a natural system that has the integrity, the independence and the benign dependence of an organism"

— Wendell Berry, "The Gift of Good Land"

The use of "organic" popularized by Howard and Rodale, on the other hand, refers more narrowly to the use of organic matter derived from plant compost and animal manures to improve the humus content of soils, grounded in the work of early soil scientists who developed what was then called "humus farming." Since the early 1940s the two camps have tended to merge.

Methods

Organic cultivation of mixed vegetables in Capay, California. Note the hedgerow in the background.

"Organic agriculture is a production system that sustains the health of soils, ecosystems and people. It relies on ecological processes, biodiversity and cycles adapted to local conditions, rather than the use of inputs with adverse effects. Organic agriculture combines tradition, innovation and science to benefit the shared environment and promote fair relationships and a good quality of life for all involved..."

— International Federation of Organic Agriculture Movements

Organic farming methods combine scientific knowledge of ecology and modern technology with traditional farming practices based on naturally occurring biological processes. Organic farming methods are studied in the field of agroecology. While conventional agriculture uses synthetic pesticides and water-soluble synthetically purified fertilizers, organic farmers are restricted by regulations to using natural pesticides and fertilizers. An example of a natural pesticide is pyrethrin, which is found naturally in the Chrysanthemum flower. The principal methods of organic farming include crop rotation, green manures and compost, biological pest control, and mechanical cultivation. These measures use the natural environment to enhance agricultural productivity: legumes are planted to fix nitrogen into the soil, natural insect predators are encouraged, crops are rotated to confuse pests and renew soil, and natural materials such as potassium bicarbonate and mulches are used to control disease and weeds. Genetically modified seeds and animals are excluded.

While organic is fundamentally different from conventional because of the use of carbon based fertilizers compared with highly soluble synthetic based fertilizers and biological pest control instead of synthetic pesticides, organic farming and large-scale conventional farming are not entirely mutually exclusive. Many of the methods developed for organic agriculture have been borrowed by more conventional agriculture. For example, Integrated Pest Management is a multifaceted strategy that uses various organic methods of pest control whenever possible, but in conventional farming could include synthetic pesticides only as a last resort.

Crop Diversity

Organic farming encourages Crop diversity. The science of agroecology has revealed the benefits of polyculture (multiple crops in the same space), which is often employed in organic farming. Planting a variety of vegetable crops supports a wider range of beneficial insects, soil microorganisms, and other factors that add up to overall farm health. Crop diversity helps environments thrive and protects species from going extinct.

Soil Management

Organic farming relies heavily on the natural breakdown of organic matter, using techniques like green manure and composting, to replace nutrients taken from the soil by previous crops. This biological process, driven by microorganisms such as mycorrhiza, allows the natural production of nutrients in the soil throughout the growing season, and has been referred to as *feeding the soil to feed the plant*. Organic farming uses a variety of methods to improve soil fertility, including crop rotation, cover cropping, reduced tillage, and application of compost. By reducing tillage, soil is not inverted and exposed to air; less carbon is lost to the atmosphere resulting in more soil organic carbon. This has an added benefit of carbon sequestration, which can reduce green house gases and help reverse climate change.

Plants need nitrogen, phosphorus, and potassium, as well as micronutrients and symbiotic relationships with fungi and other organisms to flourish, but getting enough nitrogen, and particularly synchronization so that plants get enough nitrogen at the right time (when plants need it most), is a challenge for organic farmers. Crop rotation and green manure ("cover crops") help to provide nitrogen through legumes (more precisely, the *Fabaceae* family), which fix nitrogen from the atmosphere through symbiosis with rhizobial bacteria. Intercropping, which is sometimes used for insect and disease control, can also increase soil nutrients, but the competition between the legume and the crop can be problematic and wider spacing between crop rows is required. Crop residues can be ploughed back into the soil, and different plants leave different amounts of nitrogen, potentially aiding synchronization. Organic farmers also use animal manure, certain processed fertilizers such as seed meal and various mineral powders such as rock phosphate and green sand, a naturally occurring form of potash that provides potassium. Together these methods help to control erosion. In some cases pH may need to be amended. Natural pH amendments include lime and sulfur, but in the U.S. some compounds such as iron sulfate, aluminum sulfate, magnesium sulfate, and soluble boron products are allowed in organic farming.

Mixed farms with both livestock and crops can operate as ley farms, whereby the land gathers fertility through growing nitrogen-fixing forage grasses such as white clover or alfalfa and grows cash crops or cereals when fertility is established. Farms without livestock ("stockless") may find it more difficult to maintain soil fertility, and may rely more on external inputs such as imported manure as well as grain legumes and green manures, although grain legumes may fix limited nitrogen because they are harvested. Horticultural farms that grow fruits and vegetables in protected conditions often relay even more on external inputs.

Biological research into soil and soil organisms has proven beneficial to organic farming. Varieties of bacteria and fungi break down chemicals, plant matter and animal waste into productive soil nutrients. In turn, they produce benefits of healthier yields and more productive soil for future crops. Fields with less or no manure display significantly lower yields, due to decreased soil microbe community. Increased manure improves biological activity, providing a healthier, more arable soil system and higher yields.

Weed Management

Organic weed management promotes weed suppression, rather than weed elimination, by enhancing crop competition and phytotoxic effects on weeds. Organic farmers integrate cultural, biological, mechanical, physical and chemical tactics to manage weeds without synthetic herbicides.

Organic standards require rotation of annual crops, meaning that a single crop cannot be grown in the same location without a different, intervening crop. Organic crop rotations frequently include weed-suppressive cover crops and crops with dissimilar life cycles to discourage weeds associated with a particular crop. Research is ongoing to develop organic methods to promote the growth of natural microorganisms that suppress the growth or germination of common weeds.

Other cultural practices used to enhance crop competitiveness and reduce weed pressure include selection of competitive crop varieties, high-density planting, tight row spacing, and late planting into warm soil to encourage rapid crop germination.

Mechanical and physical weed control practices used on organic farms can be broadly grouped as:

- Tillage - Turning the soil between crops to incorporate crop residues and soil amendments; remove existing weed growth and prepare a seedbed for planting; turning soil after seeding to kill weeds, including cultivation of row crops;

- Mowing and cutting - Removing top growth of weeds;

- Flame weeding and thermal weeding - Using heat to kill weeds; and

- Mulching - Blocking weed emergence with organic materials, plastic films, or landscape fabric.

Some critics, citing work published in 1997 by David Pimentel of Cornell University, which described an epidemic of soil erosion worldwide, have raised concerned that tillage contribute to the erosion epidemic. The FAO and other organizations have advocated a 'no-till' approach to both conventional and organic farming, and point out in particular that crop rotation techniques used in organic farming are excellent no-till approaches. A study published in 2005 by Pimentel and colleagues confirmed that 'Crop rotations and cover cropping (green manure) typical of organic agriculture reduce soil erosion, pest problems, and pesticide use.' Some naturally sourced chemicals are allowed for herbicidal use. These include certain formulations of acetic acid (concentrated vinegar), corn gluten meal, and essential oils. A few selective bioherbicides based on fungal pathogens have also been developed. At this time, however, organic herbicides and bioherbicides play a minor role in the organic weed control toolbox.

Weeds can be controlled by grazing. For example, geese have been used successfully to weed a range of organic crops including cotton, strawberries, tobacco, and corn, reviving the practice of keeping cotton patch geese, common in the southern U.S. before the 1950s. Similarly, some rice farmers introduce ducks and fish to wet paddy fields to eat both weeds and insects.

Controlling other Organisms

Organisms aside from weeds that cause problems on organic farms include arthropods (e.g., insects, mites), nematodes, fungi and bacteria. Organic practices include, but are not limited to:

- encouraging predatory beneficial insects to control pests by serving them nursery plants and/or an alternative habitat, usually in a form of a shelterbelt, hedgerow, or beetle bank;

- encouraging beneficial microorganisms;

- rotating crops to different locations from year to year to interrupt pest reproduction cycles;

- planting companion crops and pest-repelling plants that discourage or divert pests;

- using row covers to protect crops during pest migration periods;

- using biologic pesticides and herbicides

- using stale seed beds to germinate and destroy weeds before planting

- using sanitation to remove pest habitat;

- Using insect traps to monitor and control insect populations.

- Using physical barriers, such as row covers

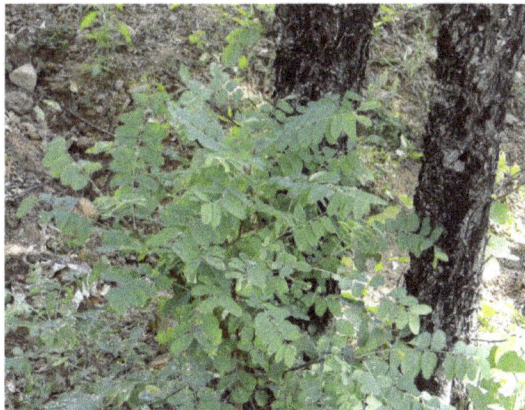

Chloroxylon is used for Pest Management in Organic Rice Cultivation in Chhattisgarh, India

Examples of predatory beneficial insects include minute pirate bugs, big-eyed bugs, and to a lesser extent ladybugs (which tend to fly away), all of which eat a wide range of pests. Lacewings are also effective, but tend to fly away. Praying mantis tend to move more slowly and eat less heavily. Parasitoid wasps tend to be effective for their selected prey, but like all small insects can be less effective outdoors because the wind controls their movement. Predatory mites are effective for controlling other mites.

Naturally derived insecticides allowed for use on organic farms use include *Bacillus thuringiensis* (a bacterial toxin), pyrethrum (a chrysanthemum extract), spinosad (a bacterial metabolite), neem (a tree extract) and rotenone (a legume root extract). Fewer than 10% of organic farmers use these pesticides regularly; one survey found that only 5.3% of vegetable growers in

California use rotenone while 1.7% use pyrethrum. These pesticides are not always more safe or environmentally friendly than synthetic pesticides and can cause harm. The main criterion for organic pesticides is that they are naturally derived, and some naturally derived substances have been controversial. Controversial natural pesticides include rotenone, copper, nicotine sulfate, and pyrethrums Rotenone and pyrethrum are particularly controversial because they work by attacking the nervous system, like most conventional insecticides. Rotenone is extremely toxic to fish and can induce symptoms resembling Parkinson's disease in mammals. Although pyrethrum (natural pyrethrins) is more effective against insects when used with piperonyl butoxide (which retards degradation of the pyrethrins), organic standards generally do not permit use of the latter substance.

Naturally derived fungicides allowed for use on organic farms include the bacteria *Bacillus subtilis* and *Bacillus pumilus*; and the fungus *Trichoderma harzianum*. These are mainly effective for diseases affecting roots. Compost tea contains a mix of beneficial microbes, which may attack or out-compete certain plant pathogens, but variability among formulations and preparation methods may contribute to inconsistent results or even dangerous growth of toxic microbes in compost teas.

Some naturally derived pesticides are not allowed for use on organic farms. These include nicotine sulfate, arsenic, and strychnine.

Synthetic pesticides allowed for use on organic farms include insecticidal soaps and horticultural oils for insect management; and Bordeaux mixture, copper hydroxide and sodium bicarbonate for managing fungi. Copper sulfate and Bordeaux mixture (copper sulfate plus lime), approved for organic use in various jurisdictions, can be more environmentally problematic than some synthetic fungicides dissallowed in organic farming Similar concerns apply to copper hydroxide. Repeated application of copper sulfate or copper hydroxide as a fungicide may eventually result in copper accumulation to toxic levels in soil, and admonitions to avoid excessive accumulations of copper in soil appear in various organic standards and elsewhere. Environmental concerns for several kinds of biota arise at average rates of use of such substances for some crops. In the European Union, where replacement of copper-based fungicides in organic agriculture is a policy priority, research is seeking alternatives for organic production.

Livestock

For livestock like these healthy cows vaccines play an important part in animal health since antibiotic therapy is prohibited in organic farming

Raising livestock and poultry, for meat, dairy and eggs, is another traditional farming activity that complements growing. Organic farms attempt to provide animals with natural living conditions and feed. Organic certification verifies that livestock are raised according to the USDA organic regulations throughout their lives. These regulations include the requirement that all animal feed must be certified organic.

Organic livestock may be, and must be, treated with medicine when they are sick, but drugs cannot be used to promote growth, their feed must be organic, and they must be pastured.

Also, horses and cattle were once a basic farm feature that provided labor, for hauling and plowing, fertility, through recycling of manure, and fuel, in the form of food for farmers and other animals. While today, small growing operations often do not include livestock, domesticated animals are a desirable part of the organic farming equation, especially for true sustainability, the ability of a farm to function as a self-renewing unit.

Genetic Modification

A key characteristic of organic farming is the rejection of genetically engineered plants and animals. On 19 October 1998, participants at IFOAM's 12th Scientific Conference issued the Mar del Plata Declaration, where more than 600 delegates from over 60 countries voted unanimously to exclude the use of genetically modified organisms in food production and agriculture.

Although opposition to the use of any transgenic technologies in organic farming is strong, agricultural researchers Luis Herrera-Estrella and Ariel Alvarez-Morales continue to advocate integration of transgenic technologies into organic farming as the optimal means to sustainable agriculture, particularly in the developing world, as does author and scientist Pamela Ronald, who views this kind of biotechnology as being consistent with organic principles.

Although GMOs are excluded from organic farming, there is concern that the pollen from genetically modified crops is increasingly penetrating organic and heirloom seed stocks, making it difficult, if not impossible, to keep these genomes from entering the organic food supply. Differing regulations among countries limits the availability of GMOs to certain countries, as described in the article on regulation of the release of genetic modified organisms.

Tools

Organic farmers use a number of traditional farm tools to do farming. Due to the goals of sustainability in organic farming, organic farmers try to minimize their reliance on fossil fuels. In the developing world on small organic farms tools are normally constrained to hand tools and diesel powered water pumps. A recent study evaluated the use of open-source 3-D printers (called RepRaps using a bioplastic polylactic acid (PLA) on organic farms. PLA is a strong biodegradable and recyclable thermoplastic appropriate for a range of representative products in five categories of prints: handtools, food processing, animal management, water management and hydroponics. Such open source hardware is attractive to all types of small farmers as it provides control for farmers over their own equipment; this is exemplified by Open Source Ecology, Farm Hack and FarmBot.

Standards

Standards regulate production methods and in some cases final output for organic agriculture. Standards may be voluntary or legislated. As early as the 1970s private associations certified organic producers. In the 1980s, governments began to produce organic production guidelines. In the 1990s, a trend toward legislated standards began, most notably with the 1991 EU-Eco-regulation developed for European Union, which set standards for 12 countries, and a 1993 UK program. The EU's program was followed by a Japanese program in 2001, and in 2002 the U.S. created the National Organic Program (NOP). As of 2007 over 60 countries regulate organic farming (IFOAM 2007:11). In 2005 IFOAM created the Principles of Organic Agriculture, an international guideline for certification criteria. Typically the agencies accredit certification groups rather than individual farms.

Organic production materials used in and foods are tested independently by the Organic Materials Review Institute.

Composting

Using manure as a fertiliser risks contaminating food with animal gut bacteria, including pathogenic strains of E. coli that have caused fatal poisoning from eating organic food. To combat this risk, USDA organic standards require that manure must be sterilized through high temperature thermophilic composting. If raw animal manure is used, 120 days must pass before the crop is harvested if the final product comes into direct contact with the soil. For products that don't directly contact soil, 90 days must pass prior to harvest.

Economics

The economics of organic farming, a subfield of agricultural economics, encompasses the entire process and effects of organic farming in terms of human society, including social costs, opportunity costs, unintended consequences, information asymmetries, and economies of scale. Although the scope of economics is broad, agricultural economics tends to focus on maximizing yields and efficiency at the farm level. Economics takes an anthropocentric approach to the value of the natural world: biodiversity, for example, is considered beneficial only to the extent that it is valued by people and increases profits. Some entities such as the European Union subsidize organic farming, in large part because these countries want to account for the externalities of reduced water use, reduced water contamination, reduced soil erosion, reduced carbon emissions, increased biodiversity, and assorted other benefits that result from organic farming.

Traditional organic farming is labor and knowledge-intensive whereas conventional farming is capital-intensive, requiring more energy and manufactured inputs.

Organic farmers in California have cited marketing as their greatest obstacle.

Geographic Producer Distribution

The markets for organic products are strongest in North America and Europe, which as of 2001 are estimated to have $6 and $8 billion respectively of the $20 billion global market. As of 2007 Australasia has 39% of the total organic farmland, including Australia's 1,180,000 hectares (2,900,000 acres) but 97 percent of this land is sprawling rangeland (2007:35). US sales are 20x as much. Europe farms 23 percent of global organic farmland (6,900,000 ha (17,000,000 acres)), followed by Latin America with 19 percent (5.8 million hectares - 14.3 million acres). Asia has 9.5 percent while North America has 7.2 percent. Africa has 3 percent.

Besides Australia, the countries with the most organic farmland are Argentina (3.1 million hectares - 7.7 million acres), China (2.3 million hectares - 5.7 million acres), and the United States (1.6 million hectares - 4 million acres). Much of Argentina's organic farmland is pasture, like that of Australia (2007:42). Spain, Germany, Brazil (the world's largest agricultural exporter), Uruguay, and the UK follow the United States in the amount of organic land (2007:26).

In the European Union (EU25) 3.9% of the total utilized agricultural area was used for organic production in 2005. The countries with the highest proportion of organic land were Austria (11%) and Italy (8.4%), followed by the Czech Republic and Greece (both 7.2%). The lowest figures were shown for Malta (0.1%), Poland (0.6%) and Ireland (0.8%). In 2009, the proportion of organic land in the EU grew to 4.7%. The countries with highest share of agricultural land were Liechtenstein (26.9%), Austria (18.5%) and Sweden (12.6%). 16% of all farmers in Austria produced organically in 2010. By the same year the proportion of organic land increased to 20%.: In 2005 168,000 ha (415,000 ac) of land in Poland was under organic management. In 2012, 288,261 hectares (712,308 acres) were under organic production, and there were about 15,500 organic farmers; retail sales of organic products were EUR 80 million in 2011. As of 2012 organic exports were part of the government's economic development strategy.

After the collapse of the Soviet Union in 1991, agricultural inputs that had previously been purchased from Eastern bloc countries were no longer available in Cuba, and many Cuban farms converted to organic methods out of necessity. Consequently, organic agriculture is a mainstream practice in Cuba, while it remains an alternative practice in most other countries. Cuba's organic strategy includes development of genetically modified crops; specifically corn that is resistant to the palomilla moth

Growth

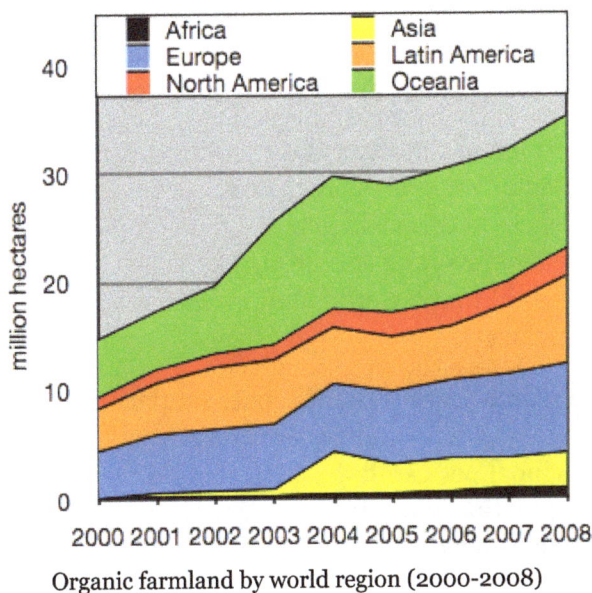

Organic farmland by world region (2000-2008)

In 2001, the global market value of certified organic products was estimated at USD $20 billion. By 2002, this was USD $23 billion and by 2015 more than USD $43 billion. By 2014, retail sales of organic products reached USD $80 billion worldwide. North America and Europe accounted for more than 90% of all organic product sales.

Organic agricultural land increased almost fourfold in 15 years, from 11 million hectares in 1999 to 43.7 million hectares in 2014. Between 2013 and 2014, organic agricultural land grew by 500,000 hectares worldwide, increasing in every region except Latin America. During this time period, Europe's organic farmland increased 260,000 hectares to 11.6 million total (+2.3%), Asia's increased 159,000 hectares to 3.6 million total (+4.7%), Africa's increased 54,000 hectares to 1.3 million total (+4.5%), and North America's increased 35,000 hectares to 3.1 million total (+1.1%). As of 2014, the country with the most organic land was Australia (17.2 million hectares), followed by Argentina (3.1 million hectares), and the United States (2.2 million hectares).

In 2013, the number of organic producers grew by almost 270,000, or more than 13%. By 2014, there were a reported 2.3 million organic producers in the world. Most of the total global increase took place in the Philippines, Peru, China, and Thailand. Overall, the majority of all organic producers are in India (650,000 in 2013), Uganda (190,552 in 2014), Mexico (169,703 in 2013) and the Philippines (165,974 in 2014).

Productivity

Studies comparing yields have had mixed results. These differences among findings can often be attributed to variations between study designs including differences in the crops studied and the methodology by which results were gathered.

A 2012 meta-analysis found that productivity is typically lower for organic farming than conventional farming, but that the size of the difference depends on context and in some cases may be very small. While organic yields can be lower than conventional yields, another meta-analysis published in Sustainable Agriculture Research in 2015, concluded that certain organic on-farm practices could help narrow this gap. Timely weed management and the application of manure in conjunction with legume forages/cover crops were shown to have positive results in increasing organic corn and soybean productivity. More experienced organic farmers were also found to have higher yields than other organic farmers who were just starting out.

Another meta-analysis published in the journal Agricultural Systems in 2011 analyzed 362 datasets and found that organic yields were on average 80% of conventional yields. The author's found that there are relative differences in this yield gap based on crop type with crops like soybeans and rice scoring higher than the 80% average and crops like wheat and potato scoring lower. Across global regions, Asia and Central Europe were found to have relatively higher yields and Northern Europe relatively lower than the average.

A 2007 study compiling research from 293 different comparisons into a single study to assess the overall efficiency of the two agricultural systems has concluded that "organic methods could produce enough food on a global per capita basis to sustain the current human population, and potentially an even larger population, without increasing the agricultural land base." The researchers also found that while in developed countries, organic systems on average produce 92% of the yield produced by conventional agriculture, organic systems produce 80% more than conventional farms in developing countries, because the materials needed for organic farming are more accessible than synthetic farming materials to farmers in some poor countries. This study was strongly contested by another study published in 2008, which stated, and was entitled, "Organic agriculture cannot feed the world" and said that the 2007 came up with "a major overestimation of the productivity of OA" "because data are misinterpreted and calculations accordingly are erroneous." Additional research needs to be conducted in the future to further clarify these claims.

Long Term Studies

A study published in 2005 compared conventional cropping, organic animal-based cropping, and organic legume-based cropping on a test farm at the Rodale Institute over 22 years. The study found that "the crop yields for corn and soybeans were similar in the organic animal, organic legume, and conventional farming systems". It also found that "significantly less fossil energy was expended to produce corn in the Rodale Institute's organic animal and organic legume systems than in the conventional production system. There was little difference in energy input between the different treatments for producing soybeans. In the organic systems, synthetic fertilizers and pesticides were generally not used". As of 2013 the Rodale study was ongoing and a thirty-year anniversary report was published by Rodale in 2012.

A long-term field study comparing organic/conventional agriculture carried out over 21 years in Switzerland concluded that "Crop yields of the organic systems averaged over 21 experimental years at 80% of the conventional ones. The fertilizer input, however, was 34 – 51% lower, indicating an efficient production. The organic farming systems used 20 – 56% less energy to produce a crop unit and per land area this difference was 36 – 53%. In spite of the considerably lower pesticide input the quality of organic products was hardly discernible from conventional analytically and even came off better in food preference trials and picture creating methods"

Profitability

In the United States, organic farming has been shown to be 2.9 to 3.8 times more profitable for the farmer than conventional farming when prevailing price premiums are taken into account. Globally, organic farming is between 22 and 35 percent more profitable for farmers than conventional methods, according to a 2015 meta-analysis of studies conducted across five continents.

The profitability of organic agriculture can be attributed to a number of factors. First, organic farmers do not rely on synthetic fertilizer and pesticide inputs, which can be costly. In addition, organic foods currently enjoy a price premium over conventionally produced foods, meaning that organic farmers can often get more for their yield.

The price premium for organic food is an important factor in the economic viability of organic farming. In 2013 there was a 100% price premium on organic vegetables and a 57% price premium for organic fruits. These percentages are based on wholesale fruit and vegetable prices, available through the United States Department of Agriculture's Economic Research Service. Price premiums exist not only for organic versus nonorganic crops, but may also vary depending on the venue where the product is sold: farmers markets, grocery stores, or wholesale to restaurants. For many producers, direct sales at farmers markets are most profitable because the farmer receives the entire markup, however this is also the most time and labor-intensive approach.

There have been signs of organic price premiums narrowing in recent years, which lowers the economic incentive for farmers to convert to or maintain organic production methods. Data from 22 years of experiments at the Rodale Institute found that, based on the current yields and production costs associated with organic farming in the United States, a price premium of only 10% is required to achieve parity with conventional farming. A separate study found that on a global scale, price premiums of only 5-7% percent were needed to break even with conventional methods. Without the price premium, profitability for farmers is mixed.

For markets and supermarkets organic food is profitable as well, and is generally sold at significantly higher prices than non-organic food.

Energy Efficiency

In the most recent assessments of the energy efficiency of organic versus conventional agriculture, results have been mixed regarding which form is more carbon efficient. Organic farm systems have more often than not been found to be more energy efficient, however, this is not always the case. More than anything, results tend to depend upon crop type and farm size.

A comprehensive comparison of energy efficiency in grain production, produce yield, and animal husbandry concluded that organic farming had a higher yield per unit of energy over the vast majority of the crops and livestock systems. For example, two studies - both comparing organically-versus conventionally-farmed apples - declare contradicting results, one saying organic farming is more energy efficient, the other saying conventionally is more efficient.

It has generally been found that the labor input per unit of yield was higher for organic systems compared with conventional production.

Sales and Marketing

Most sales are concentrated in developed nations. In 2008, 69% of Americans claimed to occasionally buy organic products, down from 73% in 2005. One theory for this change was that consumers were substituting "local" produce for "organic" produce.

Distributors

The USDA requires that distributors, manufacturers, and processors of organic products be certified by an accredited state or private agency. In 2007, there were 3,225 certified organic handlers, up from 2,790 in 2004.

Organic handlers are often small firms; 48% reported sales below $1 million annually, and 22% between $1 and $5 million per year. Smaller handlers are more likely to sell to independent natural grocery stores and natural product chains whereas large distributors more often market to natural product chains and conventional supermarkets, with a small group marketing to independent natural product stores. Some handlers work with conventional farmers to convert their land to organic with the knowledge that the farmer will have a secure sales outlet. This lowers the risk for the handler as well as the farmer. In 2004, 31% of handlers provided technical support on organic standards or production to their suppliers and 34% encouraged their suppliers to transition to organic. Smaller farms often join together in cooperatives to market their goods more effectively.

93% of organic sales are through conventional and natural food supermarkets and chains, while the remaining 7% of U.S. organic food sales occur through farmers' markets, foodservices, and other marketing channels.

Direct-to-Consumer Sales

In the 2012 Census, direct-to-consumer sales equaled $1.3 billion, up from $812 million in 2002, an increase of 60 percent. The number of farms that utilize direct-to-consumer sales was 144,530 in 2012 in comparison to 116,733 in 2002. Direct-to-consumer sales include farmers markets, community supported agriculture (CSA), on-farm stores, and roadside farm stands. Some organic farms also sell products direct to retailer, direct to restaurant and direct to institution. According to the 2008 Organic Production Survey, approximately 7% of organic farm sales went direct-to-consumers, 10% went direct to retailers, and approximately 83% went into wholesale markets. In comparison, only 0.4% of the value of convention agricultural commodities went direct-to-consumers.

While not all products sold at farmer's markets are certified organic, this direct-to-consumer avenue has become increasingly popular in local food distribution and has grown substantially since 1994. In 2014, there were 8,284 farmer's markets in comparison to 3,706 in 2004 and 1,755 in 1994, most of which are found in populated areas such as the Northeast, Midwest, and West Coast.

Labor and Employment

Organic production is more labor-intensive than conventional production. On the one hand, this increased labor cost is one factor that makes organic food more expensive. On the other hand, the increased need for labor may be seen as an "employment dividend" of organic farming, providing more jobs per unit area than conventional systems. The 2011 UNEP Green Economy Report suggests that "[a]n increase in investment in green agriculture is projected to lead to growth in employment of about 60 per cent compared with current levels" and that "green agriculture investments could create 47 million additional jobs compared with BAU2 over the next 40 years." The UNEP also argues that "[b]y greening agriculture and food distribution, more calories per person per day, more jobs and business opportunities especially in rural areas, and market-access opportunities, especially for developing countries, will be available."

World's Food Security

In 2007 the United Nations Food and Agriculture Organization (FAO) said that organic agriculture often leads to higher prices and hence a better income for farmers, so it should be promoted. However, FAO stressed that by organic farming one could not feed the current mankind, even less the bigger future population. Both data and models showed then that organic farming was far from sufficient. Therefore, chemical fertilizers were needed to avoid hunger. Other analysis by many agribusiness executives, agricultural and ecological scientists, and international agriculture experts revealed the opinion that organic farming would not only increase the world's food supply, but might be the only way to eradicate hunger.

FAO stressed that fertilizers and other chemical inputs can much increase the production, particularly in Africa where fertilizers are currently used 90% less than in Asia. For example, in Malawi the yield has been boosted using seeds and fertilizers. FAO also calls for using biotechnology, as it can help smallholder farmers to improve their income and food security.

Also NEPAD, development organization of African governments, announced that feeding Africans and preventing malnutrition requires fertilizers and enhanced seeds.

According to a more recent study in ScienceDigest, organic best management practices shows an average yield only 13% less than conventional. In the world's poorer nations where most of the world's hungry live, and where conventional agriculture's expensive inputs are not affordable by the majority of farmers, adopting organic management actually increases yields 93% on average, and could be an important part of increased food security.

Capacity Building in Developing Countries

Organic agriculture can contribute to ecologically sustainable, socio-economic development, especially in poorer countries. The application of organic principles enables employment of local resources (e.g., local seed varieties, manure, etc.) and therefore cost-effectiveness. Local and international markets for organic products show tremendous growth prospects and offer creative producers and exporters excellent opportunities to improve their income and living conditions.

Organic agriculture is knowledge intensive. Globally, capacity building efforts are underway, including localized training material, to limited effect. As of 2007, the International Federation of Organic Agriculture Movements hosted more than 170 free manuals and 75 training opportunities online.

In 2008 the United Nations Environmental Programme (UNEP) and the United Nations Conference on Trade and Development (UNCTAD) stated that "organic agriculture can be more conducive to food security in Africa than most conventional production systems, and that it is more likely to be sustainable in the long-term" and that "yields had more than doubled where organic, or near-organic practices had been used" and that soil fertility and drought resistance improved.

Millennium Development Goals

The value of organic agriculture (OA) in the achievement of the Millennium Development Goals (MDG), particularly in poverty reduction efforts in the face of climate change, is shown by its contribution to both income and non-income aspects of the MDGs. These benefits are expected to continue in the post-MDG era. A series of case studies conducted in selected areas in Asian countries by the Asian Development Bank Institute (ADBI) and published as a book compilation by ADB in Manila document these contributions to both income and non-income aspects of the MDGs. These include poverty alleviation by way of higher incomes, improved farmers' health owing to less chemical exposure, integration of sustainable principles into rural development policies, improvement of access to safe water and sanitation, and expansion of global partnership for development as small farmers are integrated in value chains.

A related ADBI study also sheds on the costs of OA programs and set them in the context of the costs of attaining the MDGs. The results show considerable variation across the case studies, suggesting that there is no clear structure to the costs of adopting OA. Costs depend on the efficiency of the OA adoption programs. The lowest cost programs were more than ten times less expensive than the highest cost ones. However, further analysis of the gains resulting from OA adoption reveals that the costs per person taken out of poverty was much lower than the estimates of the World Bank, based on income growth in general or based on the detailed costs of meeting some of the more quantifiable MDGs (e.g., education, health, and environment).

Externalities

Agriculture imposes negative externalities (uncompensated costs) upon society through public land and other public resource use, biodiversity loss, erosion, pesticides, nutrient runoff, subsidized water usage, subsidy payments and assorted other problems. Positive externalities include self-reliance, entrepreneurship, respect for nature, and air quality. Organic methods reduce some of these costs. In 2000 uncompensated costs for 1996 reached 2,343 million British pounds or £208 per ha (£84.20/ac). A study of practices in the USA published in 2005 concluded that cropland costs the economy approximately 5 to 16 billion dollars ($30–96/ha - $12–39/ac), while livestock production costs 714 million dollars. Both studies recommended reducing externalities. The 2000 review included reported pesticide poisonings but did not include speculative chronic health effects of pesticides, and the 2004 review relied on a 1992 estimate of the total impact of pesticides.

It has been proposed that organic agriculture can reduce the level of some negative externalities from (conventional) agriculture. Whether the benefits are private or public depends upon the division of property rights.

Several surveys and studies have attempted to examine and compare conventional and organic systems of farming and have found that organic techniques, while not without harm, are less damaging than conventional ones because they reduce levels of biodiversity less than conventional systems do and use less energy and produce less waste when calculated per unit area.

A 2003 to 2005 investigation by the Cranfield University for the Department for Environment Food and Rural Affairs in the UK found that it is difficult to compare the Global Warming Potential (GWP), acidification and eutrophication emissions but "Organic production often results in increased burdens, from factors such as N leaching and N2O emissions", even though primary energy use was less for most organic products. N_2O is always the largest GWP contributor except in tomatoes. However, "organic tomatoes always incur more burdens (except pesticide use)". Some emissions were lower "per area", but organic farming always required 65 to 200% more field area than non-organic farming. The numbers were highest for bread wheat (200+ % more) and potatoes (160% more).

The situation was shown dramatically in a comparison of a modern dairy farm in Wisconsin with one in New Zealand in which the animals grazed extensively. Using total farm emissions per kg milk produced as a parameter, the researchers showed that production of methane from belching was higher in the New Zealand farm, while carbon dioxide production was higher in the Wisconsin farm. Output of nitrous oxide, a gas with an estimated global warming potential 310 times that of carbon dioxide was also higher in the New Zealand farm. Methane from manure handling was similar in the two types of farm. The explanation for the finding relates to the different diets used on these farms, being based more completely on forage (and hence more fibrous) in New Zealand and containing less concentrate than in Wisconsin. Fibrous diets promote a higher proportion of acetate in the gut of ruminant animals, resulting in a higher production of methane that must be released by belching. When cattle are given a diet containing some concentrates (such as corn and soybean meal) in addition to grass and silage, the pattern of ruminal fermentation alters from acetate to mainly propionate. As a result, methane production is reduced. Capper et al. compared the environmental impact of US dairy production in 1944 and 2007. They calculated that the carbon "footprint" per billion kg (2.2 billion lb) of milk produced in 2007 was 37 percent that of equivalent milk production in 1944.

Environmental Impact and Emissions

Researchers at Oxford university analyzed 71 peer-reviewed studies and observed that organic products are sometimes worse for the environment. Organic milk, cereals, and pork generated higher greenhouse gas emissions per product than conventional ones but organic beef and olives had lower emissions in most studies. Usually organic products required less energy, but more land. Per unit of product, organic produce generates higher nitrogen leaching, nitrous oxide emissions, ammonia emissions, eutrophication and acidification potential than when conventionally grown. Other differences were not significant. The researchers concluded, as there is not singular way of doing conventional or organic farming, that the debate should go beyond the conventional vs organic debate, and more about finding specific solutions to specific circumstances.

Proponents of organic farming have claimed that organic agriculture emphasizes closed nutrient cycles, biodiversity, and effective soil management providing the capacity to mitigate and even reverse the effects of climate change and that organic agriculture can decrease fossil fuel emissions. "The carbon sequestration efficiency of organic systems in temperate climates is almost double (575-700 kg carbon per ha per year - 510-625 lb/ac/an) that of conventional treatment of soils, mainly owing to the use of grass clovers for feed and of cover crops in organic rotations."

Critics of organic farming methods believe that the increased land needed to farm organic food could potentially destroy the rainforests and wipe out many ecosystems.

Nutrient Leaching

According to the meta-analysis of 71 studies, nitrogen leaching, nitrous oxide emissions, ammonia emissions, eutrophication potential and acidification potential were higher for organic products, although in one study "nitrate leaching was 4.4-5.6 times higher in conventional plots than organic plots".

Excess nutrients in lakes, rivers, and groundwater can cause algal blooms, eutrophication, and subsequent dead zones. In addition, nitrates are harmful to aquatic organisms by themselves.

Land Use

The Oxford meta-analysis of 71 studies proved that organic farming requires 84% more land, mainly due to lack of nutrients but sometimes due to weeds, diseases or pests, lower yielding animals and land required for fertility building crops. While organic farming does not necessarily save land for wildlife habitats and forestry in all cases, the most modern breakthroughs in organic are addressing these issues with success.

Professor Wolfgang Branscheid says that organic animal production is not good for the environment, because organic chicken requires doubly as much land as "conventional" chicken and organic pork a quarter more. According to a calculation by Hudson Institute, organic beef requires triply as much land. On the other hand, certain organic methods of animal husbandry have been shown to restore desertified, marginal, and/or otherwise unavailable land to agricultural productivity and wildlife. Or by getting both forage and cash crop production from the same fields simultaneously, reduce net land use.

In England organic farming yields 55% of normal yields. While in other regions of the world, organic methods have started producing record yields.

Pesticides

A sign outside of an organic apple orchard in Pateros, Washington
reminding orchardists not to spray pesticides on these trees.

In organic farming synthetic pesticides are generally prohibited. A chemical is said to be synthetic if it does not already exist in the natural world. But the organic label goes further and usually prohibit compounds that exist in nature if they are produced by Chemical synthesis. So the prohibition is also about the method of production and not only the nature of the compound.

An non exhaustive list of organic approved pesticides with theirs Median lethal dose

- Copper(II) sulfate is used as a fungicide and is also used in conventional agriculture (LD_{50} 300 mg/kg). Conventional agriculture has the option to use the less toxic Mancozeb (LD_{50} 4,500 to 11,200 mg/kg)

- Boric acid is used as stomach poison that target insects (LD_{50}: 2660 mg/kg).

- Pyrethrin comes from chemicals extracted from flowers of the genus Pyrethrum (LD_{50} of 370 mg/kg). Its potent toxicity is used to control insects.

- Lime sulphur (aka calcium polysulfide) and sulfur are considered to be allowed, synthetic

materials (LD_{50}: 820 mg/kg)

- Rotenone is a powerful insecticide that was used to control insects (LD_{50}: 132 mg/kg). Despite the high toxicity of Rotenone to aquatic life and some links to Parkinson disease the compound is still allowed in organic farming as it is a naturally occurring compound.

- Bromomethane is a gas that is still used in the nurseries of Strawberry organic farming

- Azadirachtin is a wide spectrum very potent insecticide. Almost non toxic to mammals (LD_{50} in rats is > 3,540 mg/kg) but affects beneficial insects.

Food Quality and Safety

While there may be some differences in the amounts of nutrients and anti-nutrients when organically produced food and conventionally produced food are compared, the variable nature of food production and handling makes it difficult to generalize results, and there is insufficient evidence to make claims that organic food is safer or healthier than conventional food. Claims that organic food tastes better are not supported by evidence.

Soil Conservation

Supporters claim that organically managed soil has a higher quality and higher water retention. This may help increase yields for organic farms in drought years. Organic farming can build up soil organic matter better than conventional no-till farming, which suggests long-term yield benefits from organic farming. An 18-year study of organic methods on nutrient-depleted soil concluded that conventional methods were superior for soil fertility and yield for nutrient-depleted soils in cold-temperate climates, arguing that much of the benefit from organic farming derives from imported materials that could not be regarded as self-sustaining.

In *Dirt: The Erosion of Civilizations*, geomorphologist David Montgomery outlines a coming crisis from soil erosion. Agriculture relies on roughly one meter of topsoil, and that is being depleted ten times faster than it is being replaced. No-till farming, which some claim depends upon pesticides, is one way to minimize erosion. However, a 2007 study by the USDA's Agricultural Research Service has found that manure applications in tilled organic farming are better at building up the soil than no-till.

Biodiversity

The conservation of natural resources and biodiversity is a core principle of organic production. Three broad management practices (prohibition/reduced use of chemical pesticides and inorganic fertilizers; sympathetic management of non-cropped habitats; and preservation of mixed farming) that are largely intrinsic (but not exclusive) to organic farming are particularly beneficial for farmland wildlife. Using practices that attract or introduce beneficial insects, provide habitat for birds and mammals, and provide conditions that increase soil biotic diversity serve to supply vital ecological services to organic production systems. Advantages to certified organic operations that implement these types of production practices include: 1) decreased dependence on outside fertility inputs; 2) reduced pest management costs; 3) more reliable sources of clean water; and 4) better pollination.

Nearly all non-crop, naturally occurring species observed in comparative farm land practice studies show a preference for organic farming both by abundance and diversity. An average of 30% more species inhabit organic farms. Birds, butterflies, soil microbes, beetles, earthworms, spiders, vegetation, and mammals are particularly affected. Lack of herbicides and pesticides improve biodiversity fitness and population density. Many weed species attract beneficial insects that improve soil qualities and forage on weed pests. Soil-bound organisms often benefit because of increased bacteria populations due to natural fertilizer such as manure, while experiencing reduced intake of herbicides and pesticides. Increased biodiversity, especially from beneficial soil microbes and mycorrhizae have been proposed as an explanation for the high yields experienced by some organic plots, especially in light of the differences seen in a 21-year comparison of organic and control fields.

Biodiversity from organic farming provides capital to humans. Species found in organic farms enhance sustainability by reducing human input (e.g., fertilizers, pesticides).

The USDA's Agricultural Marketing Service (AMS) published a *Federal Register* notice on 15 January 2016, announcing the National Organic Program (NOP) final guidance on Natural Resources and Biodiversity Conservation for Certified Organic Operations. Given the broad scope of natural resources which includes soil, water, wetland, woodland and wildlife, the guidance provides examples of practices that support the underlying conservation principles and demonstrate compliance with USDA organic regulations § 205.200. The final guidance provides organic certifiers and farms with examples of production practices that support conservation principles and comply with the USDA organic regulations, which require operations to maintain or improve natural resources. The final guidance also clarifies the role of certified operations (to submit an OSP to a certifier), certifiers (ensure that the OSP describes or lists practices that explain the operator's monitoring plan and practices to support natural resources and biodiversity conservation), and inspectors (onsite inspection) in the implementation and verification of these production practices.

A wide range of organisms benefit from organic farming, but it is unclear whether organic methods confer greater benefits than conventional integrated agri-environmental programs. Organic farming is often presented as a more biodiversity-friendly practice, but the generality of the beneficial effects of organic farming is debated as the effects appear often species- and context-dependent, and current research has highlighted the need to quantify the relative effects of local- and landscape-scale management on farmland biodiversity. There are four key issues when comparing the impacts on biodiversity of organic and conventional farming: (1) It remains unclear whether a holistic whole-farm approach (i.e. organic) provides greater benefits to biodiversity than carefully targeted prescriptions applied to relatively small areas of cropped and/or non-cropped habitats within conventional agriculture (i.e. agri-environment schemes); (2) Many comparative studies encounter methodological problems, limiting their ability to draw quantitative conclusions; (3) Our knowledge of the impacts of organic farming in pastoral and upland agriculture is limited; (4) There remains a pressing need for longitudinal, system-level studies in order to address these issues and to fill in the gaps in our knowledge of the impacts of organic farming, before a full appraisal of its potential role in biodiversity conservation in agroecosystems can be made.

Regional Support for Organic Farming

India

In India, states such as Sikkim and Kerala have planned to shift to fully organic cultivation by 2015 and 2016 respectively.

Organic Horticulture

Organic horticulture is the science and art of growing fruits, vegetables, flowers, or ornamental plants by following the essential principles of organic agriculture in soil building and conservation, pest management, and heirloom variety preservation.

The Latin words *hortus* (garden plant) and *cultura* (culture) together form *horticulture*, classically defined as the culture or growing of garden plants. *Horticulture* is also sometimes defined simply as "agriculture minus the plough." Instead of the plough, horticulture makes use of human labour and gardener's hand tools, although some small machine tools like rotary tillers are commonly employed now.

An organic garden on a school campus

General

Mulches, cover crops, compost, manures, vermicompost, and mineral supplements are soil-building mainstays that distinguish this type of farming from its commercial counterpart. Through attention to good healthy soil condition, it is expected that insect, fungal, or other problems that sometimes plague plants can be minimized. However, pheromone traps, insecticidal soap sprays, and other pest-control methods available to organic farmers are also utilized by organic horticulturists.

Horticulture involves five areas of study. These areas are floriculture (includes production and marketing of floral crops), landscape horticulture (includes production, marketing and maintenance of landscape plants), olericulture (includes production and marketing of vegetables), pomology (includes production and marketing of fruits), and postharvest physiology (involves maintaining quality and preventing spoilage of horticultural crops). All of these can be, and sometimes are, pursued according to the principles of organic cultivation.

Organic horticulture (or organic gardening) is based on knowledge and techniques gathered over thousands of years. In general terms, organic horticulture involves natural processes, often taking place over extended periods of time, and a sustainable, holistic approach - while chemical-based horticulture focuses on immediate, isolated effects and reductionist strategies.

Organic Gardening Systems

There are a number of formal organic gardening and farming systems that prescribe specific techniques. They tend to be more specific than, and fit within, general organic standards. Forest gardening, a fully organic food production system which dates from prehistoric times, is thought to be the world's oldest and most resilient agroecosystem.

Biodynamic farming is an approach based on the esoteric teachings of Rudolf Steiner. The Japanese farmer and writer Masanobu Fukuoka invented a no-till system for small-scale grain production that he called Natural Farming. French intensive gardening and biointensive methods and SPIN Farming (Small Plot INtensive) are all small scale gardening techniques. These techniques were brought to the United States by Alan Chadwick in the 1930s. This method has since been promoted by John Jeavons, Director of Ecology Action. A garden is more than just a means of providing food, it is a model of what is possible in a community - everyone could have a garden of some kind (container, growing box, raised bed) and produce healthy, nutritious organic food, a farmers market, a place to pass on gardening experience, and a sharing of bounty, promoting a more sustainable way of living that would encourage their local economy. A simple 4' x 8' (32 square feet) raised bed garden based on the principles of bio-intensive planting and square foot gardening uses fewer nutrients and less water, and could keep a family, or community, supplied with an abundance of healthy, nutritious organic greens, while promoting a more sustainable way of living.

Organic gardening is designed to work with the ecological systems and minimally disturb the Earth's natural balance. Because of this organic farmers have been interested in reduced-tillage methods. Conventional agriculture uses mechanical tillage, which is plowing or sowing, which is harmful to the environment. The impact of tilling in organic farming is much less of an issue. Ploughing speeds up erosion because the soil remains uncovered for a long period of time and if it has a low content of organic matter the structural stability of the soil decreases. Organic farmers use techniques such as mulching, planting cover crops, and intercropping, to maintain a soil cover throughout most of the year. The use of compost, manure mulch and other organic fertilizers yields a higher organic content of soils on organic farms and helps limit soil degradation and erosion.

Other methods can also be used to supplement an existing garden. Methods such as composting, or vermicomposting. These practices are ways of recycling organic matter into some of the best organic fertilizers and soil conditioner. Vermicompost is especially easy. The byproduct is also an excellent source of nutrients for an organic garden.

Pest Control Approaches

Differing approaches to pest control are equally notable. In chemical horticulture, a specific insecticide may be applied to quickly kill off a particular insect pest. Chemical controls can dramatically reduce pest populations in the short term, yet by unavoidably killing (or starving) natural control insects and animals, cause an increase in the pest population in the long term, thereby creating an ever increasing problem. Repeated use of insecticides and herbicides also encourages rapid natural selection of resistant insects, plants and other organisms, necessitating increased use, or requiring new, more powerful controls.

In contrast, organic horticulture tends to tolerate some pest populations while taking the long view. Organic pest control requires a thorough understanding of pest life cycles and interactions, and involves the cumulative effect of many techniques, including:

- Allowing for an acceptable level of pest damage

- Encouraging predatory beneficial insects to flourish and eat pests

- Encouraging beneficial microorganisms

- Careful plant selection, choosing disease-resistant varieties

- Planting companion crops that discourage or divert pests

- Using row covers to protect crop plants during pest migration periods

- Rotating crops to different locations from year to year to interrupt pest reproduction cycles

- Using insect traps to monitor and control insect populations

Each of these techniques also provides other benefits, such as soil protection and improvement, fertilization, pollination, water conservation and season extension. These benefits are both complementary and cumulative in overall effect on site health. Organic pest control and biological pest control can be used as part of integrated pest management (IPM). However, IPM can include the use of chemical pesticides that are not part of organic or biological techniques.

Impact on the Global Food Supply

One controversy associated with organic food production is the matter of food produced per acre. Even with good organic practices, organic agriculture may be five to twenty-five percent less productive than conventional agriculture, depending on the crop.

Much of the productivity advantage of conventional agriculture is associated with the use of nitrogen fertilizer. However, the use, and especially the overuse, of nitrogen fertilizer has negative effects such as nitrogen runoff harming natural water supplies and increased global warming.

Organic methods have other advantages, such as healthier soil, that may make organic farming more resilient, and therefore more reliable in producing food, in the face of challenges such as climate change.

As well, world hunger is not primarily an issue of agricultural yields, but distribution and waste.

Natural Farming

Natural farming is an ecological farming approach established by Masanobu Fukuoka (1913–2008), a Japanese farmer and philosopher, introduced in his 1975 book *The One-Straw Revolution*. Fukuoka described his way of farming as 自然農法 (*shizen nōhō*) in Japanese. It is also referred to as "the Fukuoka Method", "the natural way of farming" or "do-nothing farming". The title refers not to lack of effort, but to the avoidance of manufactured inputs and equipment. Natural farming is related to fertility farming, organic farming, sustainable agriculture, agroforestry, ecoagriculture and permaculture but should be distinguished from biodynamic agriculture.

Masanobu Fukuoka, originator of the natural farming method

The system exploits the complexity of living organisms that shape each particular ecosystem. Fukuoka saw farming both as a means of producing food and as an aesthetic or spiritual approach to life, the ultimate goal of which was, "the cultivation and perfection of human beings". He suggested that farmers could benefit from closely observing local conditions. Natural farming is a closed system, one that demands no human-supplied inputs and mimics nature.

Fukuoka's ideas challenged conventions that are core to modern agro-industries, instead promoting an approach that takes advantage of the local environment. Natural farming differs from conventional organic farming, which Fukuoka considered to be another modern technique that disturbs nature.

Fukuoka claimed that his approach prevents water pollution, biodiversity loss and soil erosion, while providing ample amounts of food.

Principles

Fukuoka distilled natural farming into five principles:

1. No tillage

2. No fertilizer

3. No pesticides or herbicides

4. No weeding

5. No pruning

Though many of his plant varieties and practices relate specifically to Japan and even to local conditions in subtropical western Shikoku, his philosophy and the governing principles of his farming systems have been applied from Africa to the temperate northern hemisphere. In India, natural farming is often referred to as "Rishi Kheti". In India natural farming or rishi kheti includes ancient vedic principles of farming including use of animal waste and herbs for controlling pests and promoting growth. The rishi 's or Indian sages use cow products like buttermilk, milk, curd and its waste urine for preparing growth promoters. The Rishi or Vedic farming is regarded as non -violent farming without any usage of chemical fertilizer and pesticides. They obtain high quality natural or organic produce having medicinal values.Today still small number of farmers in Madhya Pradesh, Punjab, Maharashtra and Andhra Pradesh, Tamil Nadu use this farming in India.

Principally, natural farming minimises human labour and adopts, as closely as practical, nature's production of foods such as rice, barley, daikon or citrus in biodiverse agricultural ecosystems. Without plowing, seeds germinate well on the surface if site conditions meet the needs of the seeds placed there. Fukuoka used the presence of spiders in his fields as a key performance indicator of sustainability.}

Fukuoka specifies that the ground remain covered by weeds, white clover, alfalfa, herbaceous legumes, and sometimes deliberately sown herbaceous plants. Ground cover is present along with grain, vegetable crops and orchards. Chickens run free in orchards and ducks and carp populate rice fields.

Periodically ground layer plants including weeds may be cut and left on the surface, returning their nutrients to the soil, while suppressing weed growth. This also facilitates the sowing of seeds in the same area because the dense ground layer hides the seeds from animals such as birds.

For summer rice and winter barley grain crops, ground cover enhances nitrogen fixation. Straw from the previous crop mulches the topsoil. Each grain crop is sown before the previous one is harvested by broadcasting the seed among the standing crop. Later, this method was reduced to a single direct seeding of clover, barley and rice over the standing heads of rice. The result is a denser crop of smaller, but highly productive and stronger plants.

Fukuoka's practice and philosophy emphasised small scale operation and challenged the need for mechanised farming techniques for high productivity, efficiency and economies of scale. While his family's farm was larger than the Japanese average, he used one field of grain crops as a small-scale example of his system.

Climax Ecosystems

Ladybirds consume aphids and are considered beneficial by natural farmers that apply biological control.

In ecology, climax ecosystems are mature ecosystems that have reached a high degree of stability, productivity and diversity. Natural farmers attempt to mimic those virtues, creating a comparable climax ecosystem, and employ advanced techniques such as intercropping, companion planting and integrated pest management.

No-till

Natural farming recognizes soils as a fundamental natural asset. Ancient soils possess physical and chemical attributes that render them capable of generating and supporting life abundance. It can be argued that tilling actually degrades the delicate balance of a climax soil:

1. Tilling may destroy crucial physical characteristics of a soil such as *water suction*, its ability to send moisture upwards, even during dry spells. The effect is due to pressure differences between soil areas. Furthermore, tilling most certainly destroys soil horizons and hence disrupts the established flow of nutrients. A study suggests that reduced tillage preserves the crop residues on the top of the soil, allowing organic matter to be formed more easily and hence increasing the total organic carbon and nitrogen when compared to conventional tillage. The increases in organic carbon and nitrogen increase aerobic, facultative anaerobic and anaerobic bacteria populations.

2. Tilling over-pumps oxygen to local soil residents, such as bacteria and fungi. As a result, the chemistry of the soil changes. Biological decomposition accelerates and the microbiota mass increases at the expense of other organic matter, adversely affecting most plants, including trees and vegetables. For plants to thrive a certain quantity of organic matter (around 5%) must be present in the soil.

3. Tilling uproots all the plants in the area, turning their roots into food for bacteria and fungi. This damages their ability to aerate the soil. Living roots drill millions of tiny holes in the soil and thus provide oxygen. They also create room for beneficial insects and annelids (the phylum of worms). Some types of roots contribute directly to soil fertility by funding a mutualistic relationship with certain kinds of bacteria (most famously the rhizobium) that can fix nitrogen.

Fukuoka advocated avoiding any change in the natural landscape. This idea differs significantly from some recent permaculture practice that focuses on permaculture design, which may involve the change in landscape. For example, Sepp Holzer, an Austrian permaculture farmer, advocates the creation of terraces on slopes to control soil erosion. Fukuoka avoided the creation of terraces in his farm, even though terraces were common in China and Japan in his time. Instead, he prevented soil erosion by simply growing trees and shrubs on slopes.

Fertility Farming

In 1951, Newman Turner advocated the practice of "fertility farming", a system featuring the use of a cover crop, no tillage, no chemical fertilizers, no pesticides, no weeding and no composting. Although Turner was a commercial farmer and did not practice random seeding of seed balls, his "fertility farming" principles share similarities with Fukuoka's system of natural farming. Turner also advocated a "natural method" of animal husbandry.

Nature Farming

Japanese farmer and philosopher Mokichi Okada, conceived of a "no fertilizer" farming system in the 1930s that predated Fukuoka. Okada used the same Chinese characters, which are generally translated in English as "nature farming". Agriculture researcher Hu-lian Xu claims that "nature farming" is the correct literal translation of the Japanese term.

Organic Certification

Organic certification is a certification process for producers of organic food and other organic agricultural products. In general, any business directly involved in food production can be certified, including seed suppliers, farmers, food processors, retailers and restaurants.

The National Organic Program (run by the USDA) is in charge of labeling foods organic. In order for a food to be labeled "organic" it must meet the USDA's organic standards.

Organic vegetables at a farmers' market in Argentina

Requirements vary from country to country (List of countries with organic agriculture regulation), and generally involve a set of production standards for growing, storage, processing, packaging and shipping that include:

- avoidance of synthetic chemical inputs (e.g. fertilizer, pesticides, antibiotics, food additives), irradiation, and the use of sewage sludge;

- avoidance of genetically modified seed;

- use of farmland that has been free from prohibited chemical inputs for a number of years (often, three or more);

- for livestock, adhering to specific requirements for feed, housing, and breeding;

- keeping detailed written production and sales records (audit trail);

- maintaining strict physical separation of organic products from non-certified products;

- undergoing periodic on-site inspections.

In some countries, certification is overseen by the government, and commercial use of the term *organic* is legally restricted. Certified organic producers are also subject to the same agricultural, food safety and other government regulations that apply to non-certified producers.

Certified organic foods are not necessarily pesticide-free, certain pesticides are allowed.

Purpose

Organic certification addresses a growing worldwide demand for organic food. It is intended to assure quality and prevent fraud, and to promote commerce. While such certification was not necessary in the early days of the organic movement, when small farmers would sell their produce directly at farmers' markets, as organics have grown in popularity, more and more consumers are purchasing organic food through traditional channels, such as supermarkets. As such, consumers must rely on third-party regulatory certification.

For organic producers, certification identifies suppliers of products approved for use in certified operations. For consumers, "certified organic" serves as a product assurance, similar to "low fat", "100% whole wheat", or "no artificial preservatives".

Certification is essentially aimed at regulating and facilitating the sale of organic products to consumers. Individual certification bodies have their own service marks, which can act as branding to consumers—a certifier may promote the high consumer recognition value of its logo as a marketing advantage to producers.

Methods

Third-Party

To certify a farm, the farmer is typically required to engage in a number of new activities, in addition to normal farming operations:

- Study the organic standards, which cover in specific detail what is and is not allowed for every aspect of farming, including storage, transport and sale.

- Compliance — farm facilities and production methods must comply with the standards, which may involve modifying facilities, sourcing and changing suppliers, etc.

- Documentation — extensive paperwork is required, detailing farm history and current set-up, and usually including results of soil and water tests.

- Planning — a written annual production plan must be submitted, detailing everything from seed to sale: seed sources, field and crop locations, fertilization and pest control activities, harvest methods, storage locations, etc.

- Inspection — annual on-farm inspections are required, with a physical tour, examination of records, and an oral interview.

- Fee — an annual inspection/certification fee (currently starting at $400–$2,000/year, in the US and Canada, depending on the agency and the size of the operation). There are financial assistance programs for qualifying certified operations.

- Record-keeping — written, day-to-day farming and marketing records, covering all activities, must be available for inspection at any time.

In addition, short-notice or surprise inspections can be made, and specific tests (e.g. soil, water, plant tissue) may be requested.

For first-time farm certification, the soil must meet basic requirements of being free from use of prohibited substances (synthetic chemicals, etc.) for a number of years. A conventional farm must adhere to organic standards for this period, often two to three years. This is known as being in *transition*. Transitional crops are not considered fully organic.

Certification for operations other than farms follows a similar process. The focus is on the quality of ingredients and other inputs, and processing and handling conditions. A transport company would be required to detail the use and maintenance of its vehicles, storage facilities, containers, and so forth. A restaurant would have its premises inspected and its suppliers verified as certified organic.

Participatory

Participatory Guarantee Systems (PGS) represent an alternative to third party certification, especially adapted to local markets and short supply chains. They can also complement third party certification with a private label that brings additional guarantees and transparency. PGS enable the direct participation of producers, consumers and other stakeholders in:

- the choice and definition of the standards

- the development and implementation of certification procedures

- the certification decisions

Participatory Guarantee Systems are also referred to as "participatory certification".

Alternative Certification Options

The word *organic* is central to the certification (and organic food marketing) process, and this is also questioned by some. Where organic laws exist, producers cannot use the term legally without certification. To bypass this legal requirement for certification, various alternative certification approaches, using currently undefined terms like "authentic" and "natural", are emerging. In the US, motivated by the cost and legal requirements of certification (as of Oct. 2002), the private farmer-to-farmer association, Certified Naturally Grown, offers a "non-profit alternative eco-labelling program for small farms that grow using USDA Organic methods but are not a part of the USDA Certified Organic program."

In the UK, the interests of smaller-scale growers who use "natural" growing methods are represented by the Wholesome Food Association, which issues a symbol based largely on trust and peer-to-peer inspection.

Organic Certification and the Millennium Development Goals (MDGs)

Organic certification, as well as fair trade certification, has the potential to directly and indirectly contribute to the achievement of some of the Millennium Development Goals (MDGs), which are the eight international development goals that were established following the Millennium Summit of the United Nations in 2000, with all United Nations member states committed to help achieve the MDGs by 2015. With the growth of ethical consumerism in developed countries, imports of eco-friendly and socially certified produce from the poor in developing countries have increased, which could contribute towards the achievement of the MDGs. A study by Setboonsarng (2008) reveals that organic certification substantially contributes to MDG1 (poverty and hunger) and MDG7 (environmental sustainability) by way of premium prices and better market access, among others. This study concludes that for this market-based development scheme to broaden its poverty impacts, public sector support in harmonizing standards, building up the capacity of certifiers, developing infrastructure development, and innovating alternative certification systems will be required.

International Food Standards

The body Codex Alimentarius of the Food and Agriculture Organization of the United Nations was established in November 1961. The Commission's main goals are to protect the health of consumers and ensure fair practices in the international food trade. The Codex Alimentarius is recognized by the World Trade Organization as an international reference point for the resolution of disputes concerning food safety and consumer protection. One of their goals is to provide proper food labelling (general standard, guidelines on nutrition labelling, guidelines on labelling claims).

United States

In the United States the situation is undergoing its own FDA Food Safety Modernization Act.

Regional Variations

Organic Certification	
 Australia (ACO)	 Canada
 Australia	 European Union
 France	 Germany
 United States	 Japan

In some countries, organic standards are formulated and overseen by the government. The United States, the European Union, Canada and Japan have comprehensive organic legislation, and the term "organic" may be used only by certified producers. Being able to put the word "organic" on a food product is a valuable marketing advantage in today's consumer market, but does not guarantee the product is legitimately organic. Certification is intended to protect consumers from misuse of the term, and make buying organics easy. However, the organic labeling made possible by certification itself usually requires explanation. In countries without organic laws, government guidelines may or may not exist, while certification is handled by non-profit organizations and private companies.

Internationally, equivalency negotiations are underway, and some agreements are already in place, to harmonize certification between countries, facilitating international trade. There are also international certification bodies, including members of the International Federation of Organic Agriculture Movements (IFOAM) working on harmonization efforts. Where formal agreements do not exist between countries, organic product for export is often certified by agencies from the importing countries, who may establish permanent foreign offices for this purpose. In 2011 IFOAM introduced a new program - the IFOAM Family of Standards - that attempts to simplify harmonization. The vision is to establish the use of one single global reference (the COROS) to access the quality of standards rather than focusing on bilateral agreements.

The Certcost was a research project that conducted research and prepared reports about the certification of organic food. The project was supported by the European Commission and was active from 2008-2011. The website will be available until 2016.

North America

United States of America

In the United States, "organic" is a labeling term for food or agricultural products ("food, feed or fiber") that have been produced according to USDA organic regulations, which define standards that "integrate cultural, biological, and mechanical practices that foster cycling of resources, promote ecological balance, and conserve biodiversity." USDA standards recognize four types of organic production:

- Crops: "Plants that are grown to be harvested as food, livestock feed, or fiber used to add nutrients to the field."

- Livestock: "Animals that can be used in the production of food, fiber, or feed."

- Processed/multi-ingredient products: "Items that have been handled and packaged (e.g. chopped carrots) or combined, processed, and packaged (e.g. bread or soup)."

- Wild crops: "Plants from a growing site that is not cultivated."

Organic agricultural operations should ultimately maintain or improve soil and water quality, and conserve wetlands, woodlands, and wildlife.

In the U.S., the Organic Foods Production Act of 1990 "requires the Secretary of Agriculture to establish a National List of Allowed and Prohibited Substances which identifies synthetic substances that may be used, and the nonsynthetic substances that cannot be used, in organic production and handling operations."

Also in the U.S., the Secretary of Agriculture promulgated regulations establishing the National Organic Program (NOP). The final rule was published in the Federal Register in 2000.

USDA Organic certification confirms that the farm or handling facility (whether within the United States or internationally) complies with USDA organic regulations. Farms or handling facilities can be certified by private, foreign, or State entities, whose agents are accredited by the USDA (accredited agents are listed on the USDA website). Any farm or business that grosses more than

$5,000 annually in organic sales must be certified. Farms and businesses that make less than $5,000 annually are "exempt," and must follow all the requirements as stated in the USDA regulations except for two requirements:

- Exempt operations do not need to be certified to "sell, label, or represent" their products as organic, but may not use the USDA organic seal or label their products as "certified organic." Exempt operations may pursue optional certification if they wish to use the USDA organic seal.

- Exempt operations are not required to have a system plan that documents the specific practices and substances used in the production or handling of their organic products

Exempt operations are also barred from selling their products as ingredients for use in another producer or handler's certified organic product, and may be required by buyers to sign an affidavit affirming adherence to USDA organic regulations.

Before an operation may sell, label or represent their products as "organic" (or use the USDA organic seal), it must undergo a 3-year transition period where any land used to produce raw organic commodities must be left untreated with prohibited substances.

Operations seeking certification must first submit an application for organic certification to a USDA-accredited certifying agent including the following:

- A detailed description of the operation seeking certification

- A history of substances used on the land over the prior 3 years

- A list of the organic products grown, raised, or processed

- A written "Organic System Plan (OSP)" which outlines the practices and substances intended for use during future organic production.

- Processors/handlers who are not primarily a farm (and farms with livestock and/or crops that also process products) must complete an Organic Handling Plan (OHP), and also include a product profile and label for each product

Certifying agents then review the application to confirm that the operation's practices follow USDA regulations, and schedule an inspection to verify adherence to the OSP, maintenance of records, and overall regulatory compliance

Inspection The during the site visit, the inspector observes onsite practices and compares them to the OSP, looks for any potential contamination by prohibited materials (or any risk of potential contamination), and takes soil, tissue, or product samples as needed. At farming operations, the inspector will also examine the fields, water systems, storage areas, and equipment, assess pest and weed management, check feed production, purchase records, livestock and their living conditions, and records of animal health management practices. For processing and handling facilities, the inspector evaluates the receiving, processing, and storage areas for organic ingredients and finished products, as well as assessing any potential hazards or contamination points (from "sanitation systems, pest management materials, or nonorganic processing aids"). If the facility also processes or handles nonorganic materials, the inspector will also analyze the measures in place to prevent comingling.

If the written application and operational inspection are successful, the certifying agent will issue an organic certificate to the applicant. The producer or handler must then submit an updated application and OSP, pay recertification fees to the agent, and undergo annual onsite inspections to receive recertification annually. Once certified, producers and handlers can have up to 75% of their organic certification costs reimbursed through the USDA Organic Certification Cost-Share Programs.

Federal legislation defines three levels of organic foods. Products made entirely with certified organic ingredients, methods, and processing aids can be labeled "100% organic" (including raw agricultural commodities that have been certified), while only products with at least 95% organic ingredients may be labeled "organic" (any non-organic ingredients used must fall under the exemptions of the National List). Under these two categories, no nonorganic agricultural ingredients are allowed when organic ingredients are available. Both of these categories may also display the "USDA Organic" seal, and must state the name of the certifying agent on the information panel.

A third category, containing a minimum of 70% organic ingredients, can be labeled "made with organic ingredients," but may not display the USDA Organic seal. Any remaining agricultural ingredients must be produced without excluded methods, including genetic modification, irradiation, or the application of synthetic fertilizers, sewage sludge, or biosolids. Non-agricultural ingredients used must be allowed on the National List. Organic ingredients must be marked in the ingredients list (e.g., "organic dill" or with an asterisk denoting organic status). In addition, products may also display the logo of the certification body that approved them.

Products made with less than 70% organic ingredients can not be advertised as "organic," but can list individual ingredients that are organic as such in the product's ingredient statement. Also, USDA ingredients from plants cannot be genetically modified.

Livestock feed is only eligible for labeling as "100% Organic" or "Organic."

Alcoholic products are also subject to the Alcohol and Tobacco Tax and Trade Bureau regulations. Any use of added sulfites in wine made with organic grapes means that the product is only eligible for the "made with" labeling category and therefore may not use the USDA organic seal. Wine labeled as made with other organic fruit cannot have sulfites added to it.

Organic textiles made be labeled organic and use the USDA organic seal if the finished product is certified organic and produced in full compliance with USDA organic regulations. If all of a specific fiber used in a product is certified organic, the label may state the percentage of organic fibers and identify the organic material.

Organic certification mandates that the certifying inspector must be able to complete both "trace-back" and "mass balance audits" for all ingredients and products. A trace-back audit confirms the existence of a record trail from time of purchase/production through the final sale. A mass balance audit verifies that enough organic product and ingredients have been produced or purchased to match the amount of product sold. Each ingredient and product must have an assigned lot number to ensure the existence of a proper audit trail.

Some of the earliest organizations to carry out organic certification in North America were the California Certified Organic Farmers, founded in 1973, and the voluntary standards and certification program popularized by the Rodale Press in 1972. Some retailers have their stores certified as organic handlers and processors to ensure organic compliance is maintained throughout the supply chain until delivered to consumers, such as Vitamin Cottage Natural Grocers, a 60-year-old chain based in Colorado.

Violations of USDA Organic regulations carry fines up to $11,000 per violation, and can also lead to suspension or revocation of a farm or business's organic certificate.

Once certified, USDA organic products can be exported to countries currently engaged in organic trade agreements with the U.S., including Canada, the European Union, Japan, and Taiwan, and do not require additional certification as long as the terms of the agreement are met.

Canada

In Canada, certification was implemented at the federal level on June 30, 2009. Mandatory certification is required for agricultural products represented as organic in import, export and inter-provincial trade, or that bear the federal organic logo. In Quebec, provincial legislation provides government oversight of organic certification within the province, through the Quebec Accreditation Board (*Conseil D'Accréditation Du Québec*).

Europe

Public Organic Certification

EU countries acquired comprehensive organic legislation with the implementation of the EU-Eco-regulation 1992. Supervision of certification bodies is handled on the national level. In March 2002 the European Commission issued a EU-wide label for organic food. It has been mandatory throughout the EU since July 2010. and has become compulsory after a two-year transition period.

In 2009 a new logo was chosen through a design competition and online public vote. The new logo is a green rectangle that shows twelve stars (from the European flag) placed such that they form the shape of a leaf in the wind. Unlike earlier labels no words are presented on the label lifting the requirement for translations referring to organic food certification.

The new EU organic label has been implemented since July 2010 and has replaced the old European Organic label. However, producers that have had already printed and ready to use packaging with the old label were allowed to use them in the upcoming 2 years.

The development of the EU organic label was develop based on Denmark's organic food policy and the rules behind the Danish organic food label which at the moment holds the highest rate of recognition among its users in the world respectively 98% and 90% trust the label. The current EU organic label is meant to signal to the consumer that at least 95% of the ingredients used in the processed organic food is from organic origin and 5% considered an acceptable error margin.

European Organic Farmland in 2005		
Country	**Area (ha)**	**Percent (%)**
Austria	360,369	11
Belgium	22,994	1.7
Cyprus	2	1.1
Czech Republic	254,982	7.2
Denmark	134,129	5.2
Finland	147,587	6.5
France	560,838	2
Germany	807,406	4.7
Greece	288,737	7.2
Hungary	128,576	2
Ireland	34,912	0.8
Italy	1,069,462	8.4
Latvia	118,612	7
Lithuania	64,544	2.3
Luxembourg	3,158 *	2.4
Malta	14	0.1
Netherlands	48,765	2.5
Poland	82,730 *	2.4
Portugal	233,458	6.3
Slovakia	90,206	4.8
Slovenia	23,499	4.8
Spain	807,569	3.2
Sweden	222,268	6.2
Switzerland		11
United Kingdom	608,952	3.8
EU Total	6,115,465	3.9
Source: "Eurostat press release 80/2007"		

Private Organic Certification

Besides the public organic certification regulation EU-Eco-regulation in 1992, there are various private organic certifications available:

- Demeter International is the largest certification organization for biodynamic agriculture, and is one of three predominant organic certifiers. Demeter Biodynamic Certification is used in over 50 countries to verify that biodynamic products meet international standards in production and processing. The Demeter certification program was established in 1928, and as such was the first ecological label for organically produced foods.

- Bio Suisse established in 1981 is the Swiss organic farmer umbrella organization. International activities are mainly focused on imports towards Switzerland and don't support export activities.

Czech Republic

Following private bodies certify organic produce: KEZ, o. p. s. (CZ-BIO-001), ABCert, AG (CZ-BIO-002) and BIOCONT CZ, s. r. o. (CZ-BIO-003). These bodies provide controlling of processes tied with issueing of certificate of origin. Controlling of compliancy (to (ES) no 882/2004 directive) is provided by government body ÚKZÚZ (Central Institute for Supervising and Testing in Agriculture).

France

In France, organic certification was introduced in 1985. It has established a green-white logo of "AB - agriculture biologique." The certification for the AB label fulfills the EU regulations for organic food. The certification process is overseen a public institute ("Agence française pour le développement et la promotion de l'agriculture biologique" usually shortened to "Agence bio") established in November 2001. The actual certification authorities include a number of different institutes like Aclave, Agrocert, Ecocert SA, Qualité France SA, Ulase, SGS ICS.

Germany

In Germany the national label was introduced in September 2001 following in the footsteps of the political campaign of "Agrarwende" (*agricultural major shift*) led by minister Renate Künast of the Greens party. This campaign was started after the mad-cow disease epidemic in 2000. The effects on farming are still challenged by other political parties. The national "Bio"-label in its hexagon green-black-white shape has gained wide popularity - in 2007 there were 2431 companies having certified 41708 products. The popularity of the label is extending to neighbouring countries like Austria, Switzerland and France.

In the German-speaking countries there have been older non-government organizations that had issued labels for organic food long before the advent of the EU organic food regulations. Their labels are still used widely as they significantly exceed the requirements of the EU regulations. An organic food label like "demeter" from Demeter International has been in use since 1928 and this label is still regarded as providing the highest standards for organic food in the world. Other active NGOs include Bioland (1971), Biokreis (1979), Biopark (1991), Ecoland (1997), Ecovin (1985), Gäa e.V. (1989), Naturland (1981) and Bio Suisse (1981).

Greece

In Greece, organic certification is available from eight (8) organizations approved by EU. The major of them are BIOHELLAS and the DIO.

Ireland

In Ireland, organic certification is available from the Irish Organic Farmers and Growers Association, Demeter Standards Ltd. and Organic Trust Ltd.

Switzerland

In Switzerland, products sold as organic must comply at a minimum with the Swiss organic regulation (Regulation 910.18). Higher standards are required before a product can be labelled with the Bio Suisse label.

Sweden

In Sweden, organic certification is handled by the organisation KRAV (agriculture) with members such as farmers, processors, trade and also consumer, environmental and animal welfare interests.

United Kingdom

In the United Kingdom, organic certification is handled by a number of organizations, regulated by The Department for Environment, Food and Rural Affairs (DEFRA), of which the largest are the Soil Association and Organic Farmers and Growers. UK certification bodies are required to meet the EU minimum organic standards for all member states; they may choose to certify to standards that exceed the minimums, as is the case with the Soil Association.

The farmland converted to produce certified organic food has seen a significant evolution in the EU15 countries, rising from 1.8% in 1998 to 4.1% in 2005. For the current EU25 countries however the statistics report an overall percentage of just 1.5% as of 2005. However the statistics showed a larger turnover of organic food in some countries, reaching 10% in France and 14% in Germany. In France 21% of available vegetables, fruits, milk and eggs were certified as organic. Numbers for 2010 show that 5.4% of German farmland has been converted to produce certified organic food, as has 10.4% of Swiss farmland and 11.7% of Austrian farmland. Non-EU countries have widely adopted the European certification regulations for organic food, to increase export to EU countries.

Asia and Oceania

Australia

In Australia, organic certification is performed by several organisations that are accredited by the Biosecurity section of the Department of Agriculture (Australia), formerly the Australian Quarantine and Inspection Service, under the National Standard for Organic and Biodynamic Produce. All claims about the organic status of products sold in Australia are covered under the Competition and Consumer Act 2010.

In Australia, the The Organic Federation of Australia is the peak body for the organic industry in Australia and is part of the government's Organic Consultative Committee Legislative Working Group that sets organic standards.

Department of Agriculture accreditation is a legal requirement for all organic products exported from Australia. Export Control (Organic Produce Certification) Orders are used by the Department to assess organic certifying bodies and recognise them as approved certifying organisations. Approved certifying organisations are assessed by the Department for both initial recognition and on an at least annual basis thereafter to verify compliance.

In the absence of domestic regulation, DOA accreditation also serves as a 'de facto' benchmark for certified product sold on the domestic market. Despite its size and growing share of the economy "the organic industry in Australia remains largely self-governed. There is no specific legislation for domestic organic food standardisation and labelling at the state or federal level as there is in the USA and the EU".

Australian Approved Certifying Organisations

The Department has several approved certifying organisations that manage the certification process of organic and bio-dynamic operators in Australia. These certifying organisations perform a number of functions on the Department's behalf:

- Assess organic and bio-dynamic operators to determine compliance to the National Standard for Organic and Bio-Dynamic Produce and importing country requirements.

- Issue a Quality Management Certificate (QM Certificate) to organic operators to recognise compliance to export requirements.

- Issue Organic Produce Certificates (Export Documentation) for consignments of organic and bio-dynamic produce being exported.

As of 2015, there are seven approved certifying organisations:

- AUS-QUAL Pty Ltd (AUSQUAL)

- Australian Certified Organic (ACO)

- Bio-Dynamic Research Institute (BDRI)

- NASAA Certified Organic (NCO)

- Organic Food Chain (OFC)

- Safe Food Production Queensland (SFQ)

- Tasmanian Organic-dynamic Producers (TOP)

There are 2567 certified organic businesses reported in Australia in 2014. They include 1707 primary producers, 719 processors and manufacturers, 141 wholesalers and retailers plus other operators.

Australia does not have a logo or seal to identify which products are certified organic, instead the logos of the individual certifying organisations are used.

China

In China, the organic certification are administrate by government agency named Certification and Accreditation Administration of the People's Republic of China (CNCA). While the implementation of certification works, including site checking, lab test on soil, water, product qualities are perform by China Quality Certification Center (CQC) which is an agency of Administration of Quality Supervision,Inspection and Quarantine (AQSIQ). Organic certification procedure in china are perform according to China Organic Standard GB19630.1-4—2005 which issued on year 2005. This standard had governed standard procedure for Organic certification process perform by CQC, including application, inspection, lab test procedures, certification decision, post certification administration each. The certificate issue by CQC are valid for 1 year.

There are 2 logo are currently apply by CQC for Organic Certification, including Organic Logo and Conversion to Organic Logo.

India

In India, APEDA regulates the certification of organic products as per National Standards for Organic Production. "The NPOP standards for production and accreditation system have been recognized by European Commission and Switzerland as equivalent to their country standards. Similarly, USDA has recognized NPOP conformity assessment procedures of accreditation as equivalent to that of US. With these recognitions, Indian organic products duly certified by the accredited certification bodies of India are accepted by the importing countries." Organic food products manufactured and exported from India are marked with the India Organic certification mark issued by the APEDA. APEDA has recognized 11 inspection certification bodies, some of which are branches of foreign certification bodies, others are local certification bodies.

Japan

In Japan, the Japanese Agricultural Standard (JAS) was fully implemented as law in April 2001. This was revised in November 2005 and all JAS certifiers were required to be re-accredited by the Ministry of Agriculture.

Singapore

As of 2014 the The Agri-Food & Veterinary Authority of Singapore (AVA) had no organic certification process, but instead relied on international certification bodies; it does not track local producers who claim to have gotten organic certification.

Issues

Organic certification is not without its critics. Some of the staunchest opponents of chemical-based farming and factory farming practices also oppose formal certification. They see it as a way to drive independent organic farmers out of business, and to undermine the quality of organic food. Other organizations such as the Organic Trade Association work within the organic community to foster awareness of legislative and other related issues, and enable the influence and participation of organic proponents.

Obstacles to Small Independent Producers

Originally, in the 1960s through the 1980s, the organic food industry was composed of mainly small, independent farmers, selling locally. Organic "certification" was a matter of trust, based on a direct relationship between farmer and consumer. Critics view regulatory certification as a potential barrier to entry for small producers, by burdening them with increased costs, paperwork, and bureaucracy

In China, due to government regulations, international companies wishing to market organic produce must be independently certified. It is reported that "Australian food producers are spending up to $50,000 to be certified organic by Chinese authorities to crack the burgeoning middle-class market of the Asian superpower." Whilst the certification process is described by producers "extremely difficult and very expensive", a number of organic producers have acknowledged the ultimately positive effect of gaining access to the emerging Chinese market. For example, figures from Australian organic infant formula and baby food producer Bellamy's Organic indicate export growth, to China alone, of 70 per cent per year since gaining Chinese certification in 2008, while similar producers have shown export growth of 20 per cent to 30 per cent a year following certification

Peak Australian organic certification body, Australian Certified Organic, has stated however that "many companies have baulked at risking the money because of the complex, unwieldy and expensive process to earn Chinese certification." By comparison, equivalent certification costs in Australia are less than $2,000 (AUD), with costs in the United States as low as $750 (USD) for a similarly sized business.

Manipulative Use of Regulations

Manipulation of certification regulations as a way to mislead or outright dupe the public is a very real concern. Some examples are creating exceptions (allowing non-organic inputs to be used without loss of certification status) and creative interpretation of standards to meet the letter, but not the intention, of particular rules. For example, a complaint filed with the USDA in February 2004 against Bayliss Ranch, a food ingredient producer and its certifying agent, charged that tap water had been certified organic, and advertised for use in a variety of water-based body care and food products, in order to label them "organic" under US law. Steam-distilled plant extracts, consisting mainly of tap water introduced during the distilling process, were certified organic, and promoted as an organic base that could then be used in a claim of organic content. The case was dismissed by the USDA, as the products had been actually used only in personal care products, over which the department at the time extended no labeling control. The company subsequently adjusted its marketing by removing reference to use of the extracts in food products. Several months later, the USDA extended its organic labeling to personal care products; this complaint has not been refiled.

In 2013 the Australia Consumer Competition Commission said that water can no longer be labelled as organic water because, based on organic standards, water cannot be organic and it is misleading and deceptive to label any water as such.

False Assurance of Quality

The label itself can be used to mislead many customers that foods labelled as being organic are safer, healthier, and more nutritious.

Erosion of Standards

Critics of formal certification also fear an erosion of organic standards. Provided with a legal framework within which to operate, lobbyists can push for amendments and exceptions favorable to large-scale production, resulting in "legally organic" products produced in ways similar to current conventional food. Combined with the fact that organic products are now sold predominantly through high volume distribution channels such as supermarkets, the concern is that the market is evolving to favor the biggest producers, and this could result in the small organic farmer being squeezed out.

In the United States large food companies, have "assumed a powerful role in setting the standards for organic foods." Many members of standard-setting boards come from large food corporations. As more corporate members have joined, many nonorganic substances have been added to the National List of acceptable ingredients. The United States Congress has also played a role in allowing exceptions to organic food standards. In December 2005, the 2006 agricultural appropriations bill was passed with a rider allowing 38 synthetic ingredients to be used in organic foods, including food colorings, starches, sausage and hot-dog casings, hops, fish oil, chipotle chili pepper, and gelatin; this allowed Anheuser-Busch in 2007 to have its Wild Hop Lager certified organic "even though [it] uses hops grown with chemical fertilizers and sprayed with pesticides."

Organic Fertilizer

Organic fertilizers are fertilizers derived from animal matter, human excreta or vegetable matter. (e.g. compost, manure). Naturally occurring organic fertilizers include animal wastes from meat processing, peat, manure, slurry, and guano.

A cement reservoir containing cow manure mixed with water. This is common in rural Hainan Province, China. Note the bucket on a stick that the farmer uses to apply the mixture.

In contrast, the majority of fertilizers used in commercial farming are extracted from minerals (e.g., phosphate rock) or produced industrially (e.g., ammonia).

Compost bin for small-scale production of organic fertilizer

A large commercial compost operation

Examples and Sources

The main organic fertilizers are, peat, animal wastes (often from slaughter houses), plant wastes from agriculture, and treated sewage sludge.

Mineral

The main source of organic fertilizer is peat, an immature precursor to coal. Peat itself offers no nutritional value to the plants, but improves the soil by aeration and absorbing water.

Peat is the most widely used organic fertilizer.

Mined powdered limestone, rock phosphate, and Chilean saltpeter are inorganic (not of biologic origins) compounds, which can be energetically intensive to harvest.

Animal Sources

These materials include the products of the slaughter of animals. Bloodmeal, bone meal, hides, hoofs, and horns are typical precursors. fish meal, and feather meal are other sources.

Chicken litter, which consists of chicken manure mixed with sawdust, is an organic fertilizer that has been shown to better condition soil for harvest than synthesized fertilizer. Researchers at the Agricultural Research Service (ARS) studied the effects of using chicken litter, an organic fertilizer, versus synthetic fertilizers on cotton fields, and found that fields fertilized with chicken litter had a 12% increase in cotton yields over fields fertilized with synthetic fertilizer. In addition to higher yields, researchers valued commercially sold chicken litter at a $17/ton premium (to a total valuation of $78/ton) over the traditional valuations of $61/ton due to value added as a soil conditioner.

Plant

Processed organic fertilizers include compost, humic acid, amino acids, and seaweed extracts. Other examples are natural enzyme-digested proteins. Decomposing crop residue (green manure) from prior years is another source of fertility.

Other ARS studies have found that algae used to capture nitrogen and phosphorus runoff from agricultural fields can not only prevent water contamination of these nutrients, but also can be used as an organic fertilizer. ARS scientists originally developed the "algal turf scrubber" to reduce nutrient runoff and increase quality of water flowing into streams, rivers, and lakes. They found that this nutrient-rich algae, once dried, can be applied to cucumber and corn seedlings and result in growth comparable to that seen using synthetic fertilizers.

Treated Sewage Sludge

Although night soil (from human excreta) was a traditional organic fertilizer, the main source of this type is nowadays treated sewage sludge, also known as biosolids.

Biosolids as soil amendment is only available to less than 1% of US agricultural land. Industrial pollutants in sewage sludge prevents recycling it as fertilizer. The USDA prohibits use of sewage sludge in organic agricultural operations in the U.S. due to industrial pollution, pharmaceuticals, hormones, heavy metals, and other factors. The USDA now requires 3rd-party certification of high-nitrogen liquid organic fertilizers sold in the U.S.

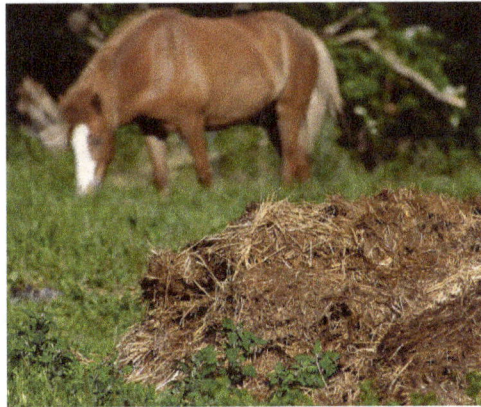

Decomposing animal manure, an organic fertilizer source

Sewage sludge use in organic agricultural operations in the U.S. has been extremely limited and rare due to USDA prohibition of the practice (due to toxic metal accumulation, among other factors).

Urine

Animal sourced urea and urea-formaldehyde from urine are suitable for organic agriculture; however, synthetically produced urea is not. The common thread that can be seen through these examples is that *organic* agriculture attempts to define itself through minimal processing, as well as being naturally occurring or via natural biological processes such as composting.

Others

- Alfalfa
- Ash
- Blood meal
- Bone meal
- Compost
- Cover crops
- Fish emulsion
- Fish meal
- Manure
- Rock phosphate
- Raw Langbeinite
- Rockdust
- Unprocessed natural potassium sulfate
- Wood chips/sawdust
- PROM

References

- Coleman, Eliot (1995), The New Organic Grower: A Master's Manual of Tools and Techniques for the Home and Market Gardener (2nd ed.), pp. 65, 108, ISBN 978-0930031756.
- Horne, Paul Anthony (2008). Integrated pest management for crops and pastures. CSIRO Publishing. p. 2. ISBN 978-0-643-09257-0.
- Vogt G (2007). Lockeretz W, ed. Chapter 1: The Origins of Organic Farming. Organic Farming: An International History. CABI Publishing. pp. 9–30. ISBN 9780851998336.
- Szykitka, Walter (2004). The Big Book of Self-Reliant Living: Advice and Information on Just About Everything You Need to Know to Live on Planet Earth. Globe-Pequot. p. 343. ISBN 978-1-59228-043-8.
- Pamela Ronald; Raoul Admachak (April 2008). "Tomorrow's Table: Organic Farming, Genetics and the Future of Food". Oxford University Press. ISBN 0195301757.
- Halberg, Niels (2006). Global development of organic agriculture: challenges and prospects. CABI. p. 297. ISBN 978-1-84593-078-3.
- Blair, Robert. (2012). Organic Production and Food Quality: A Down to Earth Analysis. Wiley-Blackwell, Oxford, UK. ISBN 978-0-8138-1217-5.

- Colin Adrien MacKinley Duncan (1996). The Centrality of Agriculture: Between Humankind and the Rest of Nature. McGill-Queen's Press - MQUP. ISBN 978-0-7735-6571-5.

- Stephen Morse; Michael Stockin (1995). People and Environment: Development for the Future. Taylor & Francis Group. ISBN 978-1-85728-283-2.

- Elpel, Thomas J. (November 1, 2002). Participating in Nature: Thomas J. Elpel's Field Guide to Primitive Living Skills. ISBN 1892784122.

- Priya Reddy; Prescott College Environmental studies (2010). Sustainable Agricultural Education: An Experiential Approach to Shifting Consciousness and Practices. Prescott College. ISBN 978-1-124-38302-6.

- Helena Norberg-Hodge; Peter Goering; John Page (1 January 2001). From the Ground Up: Rethinking Industrial Agriculture. Zed Books. ISBN 978-1-85649-994-1.

- Masanobu Fukuoka (1987). The Natural Way of Farming: The Theory and Practice of Green Philosophy. Japan Publications. ISBN 978-0-87040-613-3.

- Sylvia, D.M.; Fuhrmann, J.J.; Hartel, P.G.; Zuberer, D.A. (1999). Principles and Applications of Soil Microbiology. New Jersey: Prentice Hall. pp. 39–41. ISBN 0130941174.

6

Agricultural Productivity

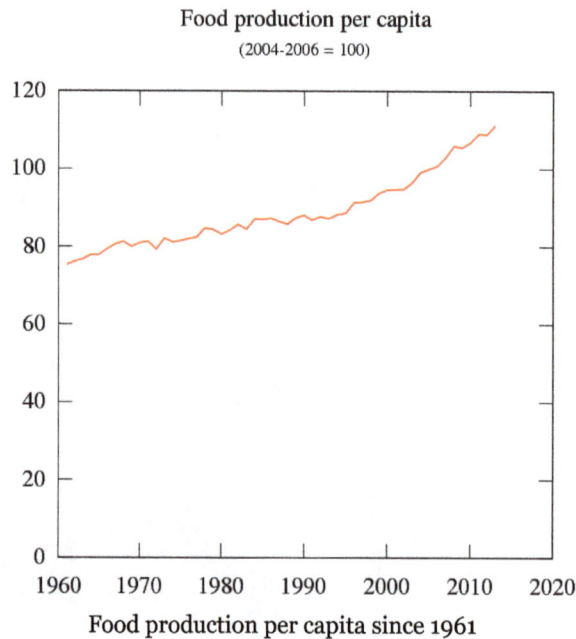

Agricultural productivity is the ratio of agriculture outputs to agricultural inputs. The major processes of agricultural productivity elucidated in this chapter are mechanised agriculture, irrigation, theoretical production ecology, leaf area index and animal feed. This chapter has been carefully written to provide an easy understanding of the varied processes of agricultural productivity.

Agricultural Productivity

Food production per capita
(2004-2006 = 100)

Food production per capita since 1961

Agricultural productivity is measured as the ratio of agricultural outputs to agricultural inputs. While individual products are usually measured by weight, their varying densities make measuring overall agricultural output difficult. Therefore, output is usually measured as the market value of final output, which excludes intermediate products such as corn feed used in the meat industry. This output value may be compared to many different types of inputs such as labour and land (yield). These are called partial measures of productivity.

Agricultural productivity may also be measured by what is termed total factor productivity (TFP). This method of calculating agricultural productivity compares an index of agricultural inputs to an index of outputs. This measure of agricultural productivity was established to remedy the shortcomings of the partial measures of productivity; notably that it is often hard to identify the factors cause them to change. Changes in TFP are usually attributed to technological improvements.

Grain production facilities

Sources of Agricultural Productivity

Wheat yields in Least Developed Countries

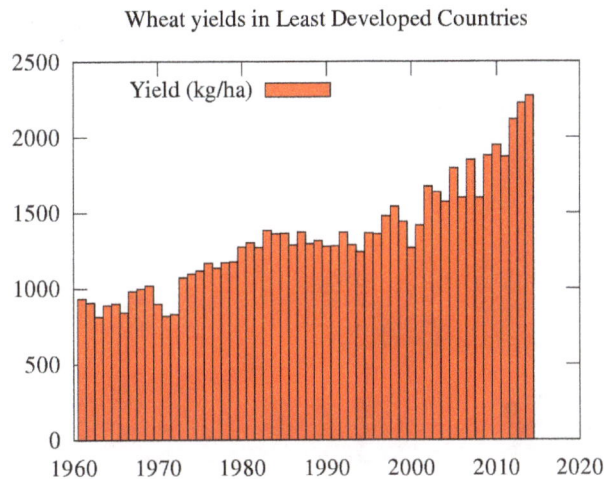

Wheat yields in least developed countries since 1961. The steep rise in crop yields in the U.S. began in the 1940s.
The percentage of growth was fastest in the early rapid growth stage. In developing countries
maize yields are still rapidly rising.

Some sources of agricultural productivity are:

- Mechanization

- High yield varieties, which were the basis of the Green revolution

- Fertilizers: Primary plant nutrients: nitrogen, phosphorus and potassium and secondary nutrients such as sulfur, zinc, copper, manganese, calcium, magnesium and molybdenum on deficient soil

- Liming of acid soils to raise pH and to provide calcium and magnesium

- Irrigation

- Herbicides

- Pesticides

- Increased plant density

- Animal feed made more digestible by processing

- Keeping animals indoors in cold weather

Importance of Agricultural Productivity

The productivity of a region's farms is important for many reasons. Aside from providing more food, increasing the productivity of farms affects the region's prospects for growth and competitiveness on the agricultural market, income distribution and savings, and labour migration. An increase in a region's agricultural productivity implies a more efficient distribution of scarce resources. As farmers adopt new techniques and differences, the more productive farmers benefit from an increase in their welfare while farmers who are not productive enough will exit the market to seek success elsewhere.

A cooperative dairy factory in Victoria.

As a region's farms become more productive, its comparative advantage in agricultural products increases, which means that it can produce these products at a lower opportunity cost than can other regions. Therefore, the region becomes more competitive on the world market, which means that it can attract more consumers since they are able to buy more of the products offered for the same amount of money.

Increases in agricultural productivity lead also to agricultural growth and can help to alleviate poverty in poor and developing countries, where agriculture often employs the greatest portion of the population. As farms become more productive, the wages earned by those who work in agriculture increase. At the same time, food prices decrease and food supplies become more stable. Labourers therefore have more money to spend on food as well as other products. This also leads to agricultural growth. People see that there is a greater opportunity to earn their living by farming and are attracted to agriculture either as owners of farms themselves or as labourers.

However, it is not only the people employed in agriculture who benefit from increases in agricultural productivity. Those employed in other sectors also enjoy lower food prices and a more stable food supply. Their wages may also increase.

A liquid manure spreader.

Agricultural productivity is becoming increasingly important as the world population continues to grow. India, one of the world's most populous countries, has taken steps in the past decades to increase its land productivity. Forty years ago, North India produced only wheat, but with the advent of the earlier maturing high-yielding wheats and rices, the wheat could be harvested in time to plant rice. This wheat/rice combination is now widely used throughout the Punjab, Haryana, and parts of Uttar Pradesh. The wheat yield of three tons and rice yield of two tons combine for five tons of grain per hectare, helping to feed India's 1.1 billion people.

Agricultural Productivity and Sustainable Development

Increase in agricultural productivity is often linked with questions about sustainability and sustainable development. Changes in agricultural practices necessarily bring changes in demands on resources. This means that as regions implement measures to increase the productivity of their farm land, they must also find ways to ensure that future generations will also have the resources they will need to live and thrive.

U.S. Agriculture Productivity

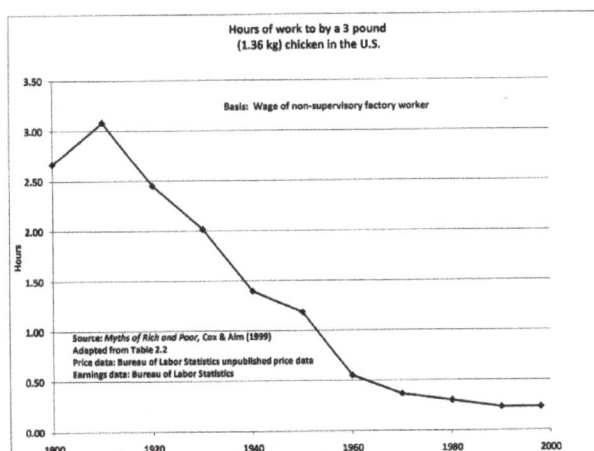

An hour's work in 1998 bought 11 times as much chicken as in 1900. Shown this way includes the increase in overall buying power plus lowered costs from the farm to the consumer.

Between 1950 and 2000, during the so-called "second agricultural revolution of modern times",

U.S. agricultural productivity rose fast, especially due to the development of new technologies. For example, the average amount of milk produced per cow increased from 5,314 pounds to 18,201 pounds per year (+242%), the average yield of corn rose from 39 bushels to 153 bushels per acre (+292%), and each farmer in 2000 produced on average 12 times as much farm output per hour worked as a farmer did in 1950.

Productive Farms

Some essential food products including bread, rice and pasta

For many farmers (especially in non-industrial countries) agricultural productivity may mean much more. A productive farm is one that provides most of the resources necessary for the farmer's family to live, such as food, fuel, fiber, healing plants, etc. It is a farm which ensures food security as well as a way to sustain the well-being of a community. This implies that a productive farm is also one which is able to ensure proper management of natural resources, such as biodiversity, soil, water, etc. For most farmers, a productive farm would also produce more goods than required for the community in order to allow trade.

Diversity in agricultural production is one key to productivity, as it enables risk management and preserves potentials for adaptation and change. Monoculture is an example of such a nondiverse production system. In a monocultural system a farmer may produce only crops, but no livestock, or only livestock and no crop.

The benefits of raising livestock, among others, are that it provides multiple goods, such as food, wool, hides, and transportation. It also has an important value in term of social relationships (such as gifts in weddings). In case of famine, when crops are not sufficient to ensure food safety, livestock can be used as food. Livestock may also provide manure, which can be used to fertilize cultivated soils, which increases soil productivity. On the other hand, in an agricultural system based only on raising livestock, food has to be bought from other farmers, and wastes produced cannot be easily disposed of. Production has many functions, and diversity is the foundation of such production. To ignore the complex functions provided by a farm is thought by many to turn agricultural production into a commodity.

Mechanised Agriculture

A cotton picker at work. The first successful models were introduced in the mid-1940s and each could do the work of 50 hand pickers.

Mechanised agriculture is the process of using agricultural machinery to mechanise the work of agriculture, greatly increasing farm worker productivity. In modern times, powered machinery has replaced many farm jobs formerly carried out by manual labour or by working animals such as oxen, horses and mules.

The entire history of agriculture contains many examples of the use of tools, such as the hoe and the plough. But the ongoing integration of machines since the Industrial Revolution has allowed farming to become much less labor-intensive.

Current mechanised agriculture includes the use of tractors, trucks, combine harvesters, countless types of farm implements, airplanes and helicopters (for aerial application), and other vehicles. Precision agriculture even uses computers in conjunction with satellite imagery and satellite navigation (GPS guidance) to increase yields.

Mechanisation was one of the large factors responsible for urbanization and industrial economies. Besides improving production efficiency, mechanisation encourages large scale production and sometimes can improve the quality of farm produce. On the other hand, it can displace unskilled farm labor and can cause environmental degradation (such as pollution, deforestation, and soil erosion), especially if it is applied shortsightedly rather than holistically.

History

A reaper at Woolbrook, New South Wales

Threshing machine in 1881. Steam engines were also used to power threshing machines.
Today both reaping and threshing are done with a combine harvester.

"Better and cheaper than horses" was the theme of many advertisements of the 1910s through 1930s.

Jethro Tull's seed drill (ca. 1701) was a mechanical seed spacing and depth placing device that increased crop yields and saved seed. It was an important factor in the British Agricultural Revolution.

Since the beginning of agriculture threshing was done by hand with a flail, requiring a great deal of labor. The threshing machine, which was invented in 1794 but not widely used for several more decades, simplified the operation and allowed the use of animal power. Before the invention of the grain cradle (ca. 1790) an able bodied laborer could reap about one quarter acre of wheat in a day using a sickle. It was estimated that for each of Cyrus McCormick's horse pulled reapers (ca. 1830s) freed up five men for military service in the U.S. Civil War. Later innovations included raking and binding machines. By 1890 two men and two horses could cut, rake and bind 20 acres of wheat per day.

In the 1880s the reaper and threshing machine were combined into the combine harvester. These machines required large teams of horses or mules to pull. Steam power was applied to threshing machines in the late 19th century. There were steam engines that moved around on wheels under their own power for supplying temporary power to stationary threshing machines. These were called *road engines,* and Henry Ford seeing one as a boy was inspired to build an automobile.

With internal combustion came the first modern tractors in the early 1900s, becoming more popular after the Fordson tractor (ca. 1917). At first reapers and combine harvesters were pulled by tractors, but in the 1930s self powered combines were developed. (*Link to a chapter on agricultural mechanisation in the 20th Century at reference*)

Advertising for motorized equipment in farm journals during this era did its best to compete against horse-drawn methods with economic arguments, extolling common themes such as that a tractor "eats only when it works", that one tractor could replace many horses, and that mechanisation could allow one man to get more work done per day than he ever had before. The horse population in the U.S. began to decline in the 1920s after the conversion of agriculture and transportation to internal combustion. Peak tractor sales in the U.S. were around 1950. In addition to saving labor, this freed up much land previously used for supporting draft animals. The greatest period of growth in agricultural productivity in the U.S. was from the 1940s to the 1970s, during which time agriculture was benefiting from internal combustion powered tractors and combine harvesters, chemical fertilizers and the green revolution.

Although farmers of corn, wheat, soy, and other commodity crops had replaced most of their workers with harvesting machines and combines enabling them to efficiently cut and gather grains, growers of produce continued to rely on human pickers to avoid the bruising of the product in order to maintain the blemish-free appearance demanded of consumers. The continuous supply of illegal workers from Latin America that were willing to harvest the crops for low wages further suppressed the need for mechanization. As the number of illegal workers has continued to decline since reaching its peak in 2007 due to increased border patrols and an improving Mexican economy, the industry is increasing the use of mechanization. Proponents argue that mechanization will boost productivity and help to maintain low food prices while farm worker advocates assert that it will eliminate jobs and will give an advantage to large growers who are able to afford the required equipment.

Current Status of Future Applications

Asparagus Harvesting

Asparagus are presently harvested by hand with labor costs at 71% of production costs and 44% of selling costs. Asparagus is a difficult crop to harvest since each spear matures at a different speed making it difficult to achieve a uniform harvest. A prototype asparagus harvesting machine - using a light-beam sensor to identify the taller spears - is expected to be available for commercial use.

Blueberry Harvesting

Mechanization of Maine's blueberry industry has reduced the number of migrant workers required from 5,000 in 2005 to 1,500 in 2015 even though production has increased from 50-60 million pounds per year in 2005 to 90 million pounds in 2015.

Chili Pepper Harvesting

As of 2014, prototype chili pepper harvesters are being tested by New Mexico State University. The New Mexico green chile crop is currently hand-picked entirely by field workers as chili pods tend to bruise easily. The first commercial application commenced in 2015. The equipment is expected to increase yield per acre and help to offset a sharp decline in acreage planted due to the lack of available labor and drought conditions.

Orange Harvesting

As of 2010, approximately 10% of the processing orange acreage in Florida is harvested mechanically. Mechanisation has progressed slowly due to the uncertainty of future economic benefits due to competition from Brazil and the transitory damage to orange trees when they are harvested.

Raisin Harvesting

As of 2007, mechanised harvesting of raisins is at 45%; however the rate has slowed due to high raisin demand and prices making the conversion away from hand labor less urgent. A new strain of grape developed by the USDA that drys on the vine and is easily harvested mechanically is expected to reduce the demand for labor.

Strawberry Harvesting

Strawberries are a high cost-high value crop with the economics supporting mechanisation. In 2005, picking and hauling costs were estimated at $594 per ton or 51% of the total grower cost. However, the delicate nature of fruit make it an unlikely candidate for mechanisation in the near future. A strawberry harvester developed by Shibuya Seiki and unveiled in Japan in 2013 is able to pick a strawberry every eight seconds. The robot identifies which strawberries are ready to pick by using three separate cameras and then once identified as ready, a mechanized arm snips the fruit free and gently places it in a basket. The robot moves on rails between the rows of strawberries which are generally contained within elevated greenhouses. The machine costs 5 million yen. A new strawberry harvester made by Agrobot that will harvest strawberries on raised, hydroponic beds using 60 robotic arms is expected to be released in 2016.

Tomato Harvesting

Mechanical harvesting of tomatoes started in 1965 and as of 2010, nearly all processing tomatoes are mechanically harvested. As of 2010, 95% of the U.S. processed tomato crop is produced in California. Although fresh market tomatoes have substantial hand harvesting costs (in 2007, the costs of hand picking and hauling were $86 per ton which is 19% of total grower cost), packing and selling costs were more of a concern (at 44% of total grower cost) making it likely that cost saving efforts would be applied there.

According to a 1977 report by the California Agrarian Action Project, during the summer of 1976 in California, many harvest machines had been equipped with a photo-electric scanner that sorted out green tomatoes among the ripe red ones using infrared lights and color sensors. It worked in lieu of 5,000 hand harvesters causing displacement of innumerable farm laborers as well as wage cuts and shorter work periods. Migrant workers were hit the hardest. To withstand the rigor of the machines, new crop varieties were bred to match the automated pickers. UC Davis Professor G.C. Hanna propagated a thick-skinned tomato called VF-145. But even still, millions were damaged with impact cracks and university breeders produced a more tougher and juiceless "square round" tomato. Small farms were of insufficient size to obtain financing to purchase the equipment and within 10 years, 85% of the state's 4,000 cannery tomato farmers were out of the business. This led to a concentrated tomato industry in California that "now packed 85% of the nation's tomato products". The monoculture fields fostered rapid pest growth, requiring the use of "more than four million pounds of pesticides each year" which greatly affected the health of the soil, the farm workers, and possibly the consumers.

Irrigation

An irrigation sprinkler watering a lawn

Irrigation canal in Osmaniye, Turkey

Irrigation is the method in which a controlled amount of water is supplied to plants at regular intervals for agriculture. It is used to assist in the growing of agricultural crops, maintenance of landscapes, and revegetation of disturbed soils in dry areas and during periods of inadequate rainfall. Additionally, irrigation also has a few other uses in crop production, which include protecting plants against frost, suppressing weed growth in grain fields and preventing soil consolidation. In contrast, agriculture that relies only on direct rainfall is referred to as rain-fed or dry land farming.

Irrigation systems are also used for dust suppression, disposal of sewage, and in mining. Irrigation is often studied together with drainage, which is the natural or artificial removal of surface and sub-surface water from a given area.

Irrigation has been a central feature of agriculture for over 5,000 years and is the product of many cultures. Historically, it was the basis for economies and societies across the globe, from Asia to the Southwestern United States.

History

Animal-powered irrigation, Upper Egypt, ca. 1846

An example of an irrigation system common on the Indian subcontinent. Artistic impression on the banks of Dal Lake, Kashmir, India

Inside a karez tunnel at Turpan, Sinkiang

ArchaMIhceal ified as evidence of irrigation where the natural rainfall was insufficient to support crops for rainfed agriculture.

Perennial irrigation was practiced in the Mesopotamian plain whereby crops were regularly watered throughout the growing season by coaxing water through a matrix of small channels formed in the field.

irrigation in Tamil Nadu (India)

Ancient Egyptians practiced *Basin irrigation* using the flooding of the Nile to inundate land plots which had been surrounded by dykes. The flood water was held until the fertile sediment had settled before the surplus was returned to the watercourse. There is evidence of the ancient Egyptian pharaoh Amenemhet III in the twelfth dynasty (about 1800 BCE) using the natural lake of the Faiyum Oasis as a reservoir to store surpluses of water for use during the dry seasons, the lake swelled annually from flooding of the Nile.

The Ancient Nubians developed a form of irrigation by using a waterwheel-like device called a *sakia*. Irrigation began in Nubia some time between the third and second millennium BCE. It largely depended upon the flood waters that would flow through the Nile River and other rivers in what is now the Sudan.

In sub-Saharan Africa irrigation reached the Niger River region cultures and civilizations by the first or second millennium BCE and was based on wet season flooding and water harvesting.

Terrace irrigation is evidenced in pre-Columbian America, early Syria, India, and China. In the Zana Valley of the Andes Mountains in Peru, archaeologists found remains of three irrigation canals radiocarbon dated from the 4th millennium BCE, the 3rd millennium BCE and the 9th century CE. These canals are the earliest record of irrigation in the New World. Traces of a canal possibly dating from the 5th millennium BCE were found under the 4th millennium canal. Sophisticated irrigation and storage systems were developed by the Indus Valley Civilization in present-day Pakistan and North India, including the reservoirs at Girnar in 3000 BCE and an early canal irrigation system from circa 2600 BCE. Large scale agriculture was practiced and an extensive network of canals was used for the purpose of irrigation.

Ancient Persia (modern day Iran) as far back as the 6th millennium BCE, where barley was grown in areas where the natural rainfall was insufficient to support such a crop. The Qanats, developed in ancient Persia in about 800 BCE, are among the oldest known irrigation methods still in use today. They are now found in Asia, the Middle East and North Africa. The system comprises a network of vertical wells and gently sloping tunnels driven into the sides of cliffs and steep hills to tap groundwater. The noria, a water wheel with clay pots around the rim powered by the flow of the stream (or by animals where the water source was still), was first brought into use at about this time, by Roman settlers in North Africa. By 150 BCE the pots were fitted with valves to allow smoother filling as they were forced into the water.

The irrigation works of ancient Sri Lanka, the earliest dating from about 300 BCE, in the reign of King Pandukabhaya and under continuous development for the next thousand years, were one of the most complex irrigation systems of the ancient world. In addition to underground canals, the Sinhalese were the first to build completely artificial reservoirs to store water. Due to their engineering superiority in this sector, they were often called 'masters of irrigation'. Most of these irrigation systems still exist undamaged up to now, in Anuradhapura and Polonnaruwa, because of the advanced and precise engineering. The system was extensively restored and further extended during the reign of King Parakrama Bahu (1153–1186 CE).

China

The oldest known hydraulic engineers of China were Sunshu Ao (6th century BCE) of the Spring and Autumn Period and Ximen Bao (5th century BCE) of the Warring States period, both of whom worked on large irrigation projects. In the Sichuan region belonging to the State of Qin of ancient China, the Dujiangyan Irrigation System was built in 256 BCE to irrigate an enormous area of farmland that today still supplies water. By the 2nd century AD, during the Han Dynasty, the Chinese also used chain pumps that lifted water from lower elevation to higher elevation. These were powered by manual foot pedal, hydraulic waterwheels, or rotating mechanical wheels pulled by oxen. The water was used for public works of providing water for urban residential quarters and palace gardens, but mostly for irrigation of farmland canals and channels in the fields.

Korea

In 15th century Korea, the world's first rain gauge, *uryanggye* (Korean:우량계), was invented in 1441. The inventor was Jang Yeong-sil, a Korean engineer of the Joseon Dynasty, under the active direction of the king, Sejong the Great. It was installed in irrigation tanks as part of a nationwide system to measure and collect rainfall for agricultural applications. With this instrument, planners and farmers could make better use of the information gathered in the survey.

North America

In North America, the Hohokam were the only culture known to rely on irrigation canals to water their crops, and their irrigation systems supported the largest population in the Southwest by AD 1300. The Hohokam constructed an assortment of simple canals combined with weirs in their various agricultural pursuits. Between the 7th and 14th centuries, they also built and maintained extensive irrigation networks along the lower Salt and middle Gila rivers that rivaled the complexity of those used in the ancient Near East, Egypt, and China. These were constructed using relatively simple excavation tools, without the benefit of advanced engineering technologies, and achieved drops of a few feet per mile, balancing erosion and siltation. The Hohokam cultivated varieties of cotton, tobacco, maize, beans and squash, as well as harvested an assortment of wild plants. Late in the Hohokam Chronological Sequence, they also used extensive dry-farming systems, primarily to grow agave for food and fiber. Their reliance on agricultural strategies based on canal irrigation, vital in their less than hospitable desert environment and arid climate, provided the basis for the aggregation of rural populations into stable urban centers.

Present Extent

Irrigation ditch in Montour County, Pennsylvania, off Strawberry Ridge Road

In the mid-20th century, the advent of diesel and electric motors led to systems that could pump groundwater out of major aquifers faster than drainage basins could refill them. This can lead to permanent loss of aquifer capacity, decreased water quality, ground subsidence, and other problems. The future of food production in such areas as the North China Plain, the Punjab, and the Great Plains of the US is threatened by this phenomenon.

At the global scale, 2,788,000 km² (689 million acres) of fertile land was equipped with irrigation infrastructure around the year 2000. About 68% of the area equipped for irrigation is located in Asia, 17% in the America, 9% in Europe, 5% in Africa and 1% in Oceania. The largest contiguous areas of high irrigation density are found:

- In Northern India and Pakistan along the Ganges and Indus rivers

- In the Hai He, Huang He and Yangtze basins in China

- Along the Nile river in Egypt and Sudan

- In the Mississippi-Missouri river basin and in parts of California

Smaller irrigation areas are spread across almost all populated parts of the world.

Only eight years later in 2008, the scale of irrigated land increased to an estimated total of 3,245,566 km² (802 million acres), which is nearly the size of India.

Types of Irrigation

Various types of irrigation techniques differ in how the water obtained from the source is distributed within the field. In general, the goal is to supply the entire field uniformly with water, so that each plant has the amount of water it needs, neither too much nor too little.

Basin flood irrigation of wheat

Irrigation of land in Punjab, Pakistan

Surface Irrigation

In *surface* (*furrow*, *flood*, or *level basin*) irrigation systems, water moves across the surface of agricultural lands, in order to wet it and infiltrate into the soil. Surface irrigation can be subdivided into furrow, *borderstrip or basin irrigation*. It is often called *flood irrigation* when the irrigation results in flooding or near flooding of the cultivated land. Historically, this has been the most common method of irrigating agricultural land and still is in most parts of the world.

Where water levels from the irrigation source permit, the levels are controlled by dikes, usually plugged by soil. This is often seen in terraced rice fields (rice paddies), where the method is used to flood or control the level of water in each distinct field. In some cases, the water is pumped, or lifted by human or animal power to the level of the land. The field water efficiency of surface irrigation is typically lower than other forms of irrigation but has the potential for efficiencies in the range of 70% - 90% under appropriate management.

Localized Irrigation

Impact sprinkler head

Localized irrigation is a system where water is distributed under low pressure through a piped network, in a pre-determined pattern, and applied as a small discharge to each plant or adjacent to it. Drip irrigation, spray or micro-sprinkler irrigation and bubbler irrigation belong to this category of irrigation methods.

Subsurface Textile Irrigation

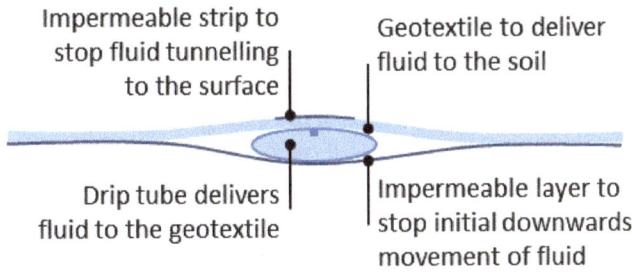

Impermeable strip to
stop fluid tunnelling
to the surface

Geotextile to deliver
fluid to the soil

Drip tube delivers
fluid to the geotextile

Impermeable layer to
stop initial downwards
movement of fluid

Diagram showing the structure of an example SSTI installation

Subsurface Textile Irrigation (SSTI) is a technology designed specifically for subsurface irrigation in all soil textures from desert sands to heavy clays. A typical subsurface textile irrigation system has an impermeable base layer (usually polyethylene or polypropylene), a drip line running along that base, a layer of geotextile on top of the drip line and, finally, a narrow impermeable layer on top of the geotextile (see diagram). Unlike standard drip irrigation, the spacing of emitters in the drip pipe is not critical as the geotextile moves the water along the fabric up to 2 m from the dripper.

Drip Irrigation

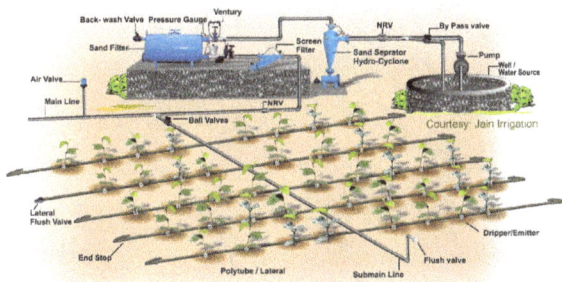

Drip irrigation layout and its parts

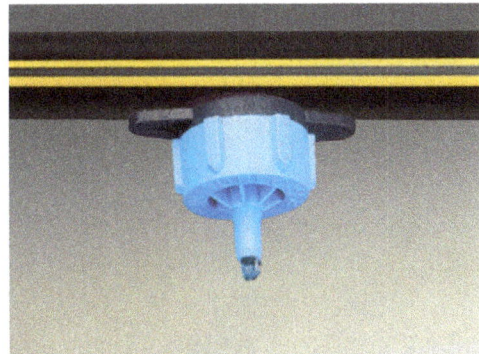

Drip irrigation – a dripper in action

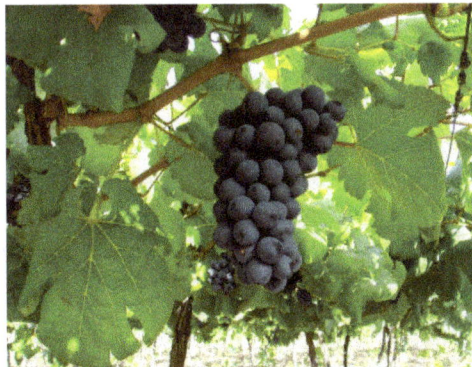

Grapes in Petrolina, only made possible in this semi arid area by drip irrigation

Drip (or micro) irrigation, also known as trickle irrigation, functions as its name suggests. In this system water falls drop by drop just at the position of roots. Water is delivered at or near the root zone of plants, drop by drop. This method can be the most water-efficient method of irrigation, if managed properly, since evaporation and runoff are minimized. The field water efficiency of drip irrigation is typically in the range of 80 to 90 percent when managed correctly.

In modern agriculture, drip irrigation is often combined with plastic mulch, further reducing evaporation, and is also the means of delivery of fertilizer. The process is known as *fertigation*.

Deep percolation, where water moves below the root zone, can occur if a drip system is operated for too long or if the delivery rate is too high. Drip irrigation methods range from very high-tech and computerized to low-tech and labor-intensive. Lower water pressures are usually needed than for most other types of systems, with the exception of low energy center pivot systems and surface irrigation systems, and the system can be designed for uniformity throughout a field or for precise water delivery to individual plants in a landscape containing a mix of plant species. Although it is difficult to regulate pressure on steep slopes, pressure compensating emitters are available, so the field does not have to be level. High-tech solutions involve precisely calibrated emitters located along lines of tubing that extend from a computerized set of valves.

Irrigation Using Sprinkler Systems

Sprinkler irrigation of blueberries in Plainville, New York, United States

A traveling sprinkler at Millets Farm Centre, Oxfordshire, United Kingdom

In *sprinkler* or overhead irrigation, water is piped to one or more central locations within the field and distributed by overhead high-pressure sprinklers or guns. A system utilizing sprinklers, sprays, or guns mounted overhead on permanently installed risers is often referred to as a *solid-set* irrigation system. Higher pressure sprinklers that rotate are called *rotors* an are driven by a ball drive, gear drive, or impact mechanism. Rotors can be designed to rotate in a full or partial circle. Guns are similar to rotors, except that they generally operate at very high pressures of 40 to 130 lbf/in² (275 to 900 kPa) and flows of 50 to 1200 US gal/min (3 to 76 L/s), usually with nozzle diameters in the range of 0.5 to 1.9 inches (10 to 50 mm). Guns are used not only for irrigation, but also for industrial applications such as dust suppression and logging.

Sprinklers can also be mounted on moving platforms connected to the water source by a hose. Automatically moving wheeled systems known as *traveling sprinklers* may irrigate areas such as small farms, sports fields, parks, pastures, and cemeteries unattended. Most of these utilize a length of polyethylene tubing wound on a steel drum. As the tubing is wound on the drum powered by the irrigation water or a small gas engine, the sprinkler is pulled across the field. When the sprinkler arrives back at the reel the system shuts off. This type of system is known to most people as a "waterreel" traveling irrigation sprinkler and they are used extensively for dust suppression, irrigation, and land application of waste water.

Other travelers use a flat rubber hose that is dragged along behind while the sprinkler platform is pulled by a cable. These cable-type travelers are definitely old technology and their use is limited in today's modern irrigation projects.

Irrigation Using Center Pivot

A small center pivot system from beginning to end

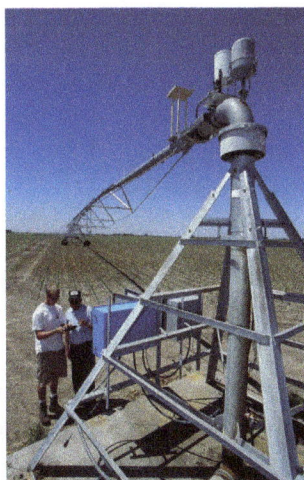

The hub of a center-pivot irrigation system

Rotator style pivot applicator sprinkler

Center pivot with drop sprinklers

Wheel line irrigation system in Idaho, 2001

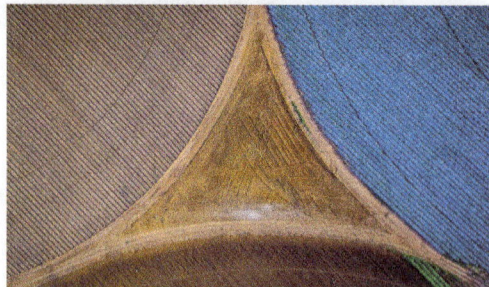
Center pivot irrigation

Center pivot irrigation is a form of sprinkler irrigation consisting of several segments of pipe (usually galvanized steel or aluminium) joined together and supported by trusses, mounted on wheeled towers with sprinklers positioned along its length. The system moves in a circular pattern and is fed with water from the pivot point at the center of the arc. These systems are found and used in all parts of the world and allow irrigation of all types of terrain. Newer systems have drop sprinkler heads as shown in the image that follows.

Most center pivot systems now have drops hanging from a u-shaped pipe attached at the top of the pipe with sprinkler head that are positioned a few feet (at most) above the crop, thus limiting evaporative losses. Drops can also be used with drag hoses or bubblers that deposit the water directly on the ground between crops. Crops are often planted in a circle to conform to the center pivot. This type of system is known as LEPA (Low Energy Precision Application). Originally, most center pivots were water powered. These were replaced by hydraulic systems (*T-L Irrigation*) and electric motor driven systems (Reinke, Valley, Zimmatic). Many modern pivots feature GPS devices.

Irrigation by Lateral Move (Side Roll, Wheel Line, Wheelmove)

A *series of pipes, each with a wheel* of about 1.5 m diameter permanently affixed to its midpoint, and sprinklers along its length, are coupled together. Water is supplied at one end using a large hose. After sufficient irrigation has been applied to one strip of the field, the hose is removed, the water drained from the system, and the assembly rolled either by hand or with a purpose-built mechanism, so that the sprinklers are moved to a different position across the field. The hose is reconnected. The process is repeated in a pattern until the whole field has been irrigated.

This system is less expensive to install than a center pivot, but much more labor-intensive to operate - it does not travel automatically across the field: it applies water in a stationary strip, must be drained, and then rolled to a new strip. Most systems use 4 or 5-inch (130 mm) diameter aluminum pipe. The pipe doubles both as water transport and as an axle for rotating all the wheels. A drive system (often found near the centre of the wheel line) rotates the clamped-together pipe sections as a single axle, rolling the whole wheel line. Manual adjustment of individual wheel positions may be necessary if the system becomes misaligned.

Wheel line systems are limited in the amount of water they can carry, and limited in the height of crops that can be irrigated. One useful feature of a lateral move system is that it consists of sections that can be easily disconnected, adapting to field shape as the line is moved. They are most often used for small, rectilinear, or oddly-shaped fields, hilly or mountainous regions, or in regions where labor is inexpensive.

Sub-irrigation

Subirrigation has been used for many years in field crops in areas with high water tables. It is a method of artificially raising the water table to allow the soil to be moistened from below the plants' root zone. Often those systems are located on permanent grasslands in lowlands or river valleys and combined with drainage infrastructure. A system of pumping stations, canals, weirs and gates allows it to increase or decrease the water level in a network of ditches and thereby control the water table.

Sub-irrigation is also used in commercial greenhouse production, usually for potted plants. Water is delivered from below, absorbed upwards, and the excess collected for recycling. Typically, a solution of water and nutrients floods a container or flows through a trough for a short period of time, 10–20 minutes, and is then pumped back into a holding tank for reuse. Sub-irrigation in greenhouses requires fairly sophisticated, expensive equipment and management. Advantages are water and nutrient conservation, and labor-saving through lowered system maintenance and automation. It is similar in principle and action to subsurface basin irrigation.

Irrigation Automatically, Non-electric Using Buckets and Ropes

Besides the common manual watering by bucket, an automated, natural version of this also exists. Using plain polyester ropes combined with a prepared ground mixture can be used to water plants from a vessel filled with water.

The ground mixture would need to be made depending on the plant itself, yet would mostly consist of black potting soil, vermiculite and perlite. This system would (with certain crops) allow to save expenses as it does not consume any electricity and only little water (unlike sprinklers, water timers, etc.). However, it may only be used with certain crops (probably mostly larger crops that do not need a humid environment; perhaps e.g. paprikas).

Irrigation using Water Condensed from Humid Air

In countries where at night, humid air sweeps the countryside.Water can be obtained from the humid air by condensation onto cold surfaces. This is for example practiced in the vineyards at Lanzarote using stones to condense water or with various fog collectors based on canvas or foil sheets.

In-ground Irrigation

Most commercial and residential irrigation systems are "in ground" systems, which means that everything is buried in the ground. With the pipes, sprinklers, emitters (drippers), and irrigation valves being hidden, it makes for a cleaner, more presentable landscape without garden hoses or other items having to be moved around manually. This does, however, create some drawbacks in the maintenance of a completely buried system.

Most irrigation systems are divided into zones. A zone is a single irrigation valve and one or a group of drippers or sprinklers that are connected by pipes or tubes. Irrigation systems are divided into zones because there is usually not enough pressure and available flow to run sprinklers for an entire yard or sports field at once. Each zone has a solenoid valve on it that is controlled via wire by an irrigation controller. The irrigation controller is either a mechanical (now the "dinosaur" type) or electrical device that signals a zone to turn on at a specific time and keeps it on for a specified amount of time. "Smart Controller" is a recent term for a controller that is capable of adjusting the watering time by itself in response to current environmental conditions. The smart controller determines current conditions by means of historic weather data for the local area, a soil moisture sensor (water potential or water content), rain sensor, or in more sophisticated systems satellite feed weather station, or a combination of these.

When a zone comes on, the water flows through the lateral lines and ultimately ends up at the irrigation emitter (drip) or sprinkler heads. Many sprinklers have pipe thread inlets on the bottom of them which allows a fitting and the pipe to be attached to them. The sprinklers are usually installed with the top of the head flush with the ground surface. When the water is pressurized, the head will pop up out of the ground and water the desired area until the valve closes and shuts off that zone. Once there is no more water pressure in the lateral line, the sprinkler head will retract back into the ground. Emitters are generally laid on the soil surface or buried a few inches to reduce evaporation losses.

Water Sources

Irrigation is underway by pump-enabled extraction directly from the Gumti, seen in the background, in Comilla, Bangladesh.

Irrigation water can come from groundwater (extracted from springs or by using wells), from surface water (withdrawn from rivers, lakes or reservoirs) or from non-conventional sources like treated wastewater, desalinated water or drainage water. A special form of irrigation using surface water is spate irrigation, also called floodwater harvesting. In case of a flood (spate), water is diverted to normally dry river beds (wadis) using a network of dams, gates and channels and spread over large areas. The moisture stored in the soil will be used thereafter to grow crops. Spate irrigation areas are in particular located in semi-arid or arid, mountainous regions. While floodwater harvesting belongs to the accepted irrigation methods, rainwater harvesting is usually not considered as a form of irrigation. Rainwater harvesting is the collection of runoff water from roofs or unused land and the concentration of this.

Around 90% of wastewater produced globally remains untreated, causing widespread water pollution, especially in low-income countries. Increasingly, agriculture uses untreated wastewater as a source of irrigation water. Cities provide lucrative markets for fresh produce, so are attractive to farmers. However, because agriculture has to compete for increasingly scarce water resources with industry and municipal users, there is often no alternative for farmers but to use water polluted with urban waste, including sewage, directly to water their crops. Significant health hazards can result from using water loaded with pathogens in this way, especially if people eat raw vegetables that have been irrigated with the polluted water. The International Water Management Institute has worked in India, Pakistan, Vietnam, Ghana, Ethiopia, Mexico and other countries on various projects aimed at assessing and reducing risks of wastewater irrigation. They advocate a 'multiple-barrier' approach to wastewater use, where farmers are encouraged to adopt various risk-reducing behaviours. These include ceasing irrigation a few days before harvesting to allow pathogens to die off in the sunlight, applying water carefully so it does not contaminate leaves likely to be eaten raw, cleaning vegetables with disinfectant or allowing fecal sludge used in farming to dry before being used as a human manure. The World Health Organization has developed guidelines for safe water use.

There are numerous benefits of using recycled water for irrigation, including the low cost (when compared to other sources, particularly in an urban area), consistency of supply (regardless of season, climatic conditions and associated water restrictions), and general consistency of quality. Irrigation of recycled wastewater is also considered as a means for plant fertilization and particularly nutrient supplementation. This approach carries with it a risk of soil and water pollution through excessive wastewater application. Hence, a detailed understanding of soil water conditions is essential for effective utilization of wastewater for irrigation.

Efficiency

Young engineers restoring and developing the old Mughal irrigation
system during the reign of the Mughal Emperor Bahadur Shah II

Modern irrigation methods are efficient enough to supply the entire field uniformly with water, so that each plant has the amount of water it needs, neither too much nor too little. Water use efficiency in the field can be determined as follows:

- Field Water Efficiency (%) = (Water Transpired by Crop ÷ Water Applied to Field) x 100

Until 1960s, the common perception was that water was an infinite resource. At that time, there were fewer than half the current number of people on the planet. People were not as wealthy as today, consumed fewer calories and ate less meat, so less water was needed to produce their food. They required a third of the volume of water we presently take from rivers. Today, the competition for water resources is much more intense. This is because there are now more than seven billion people on the planet, their consumption of water-thirsty meat and vegetables is rising, and there is increasing competition for water from industry, urbanisation and biofuel crops. To avoid a global water crisis, farmers will have to strive to increase productivity to meet growing demands for food, while industry and cities find ways to use water more efficiently.

Successful agriculture is dependent upon farmers having sufficient access to water. However, water scarcity is already a critical constraint to farming in many parts of the world. With regards to agriculture, the World Bank targets food production and water management as an increasingly global issue that is fostering a growing debate. Physical water scarcity is where there is not enough water to meet all demands, including that needed for ecosystems to function effectively. Arid regions frequently suffer from physical water scarcity. It also occurs where water seems abundant but where resources are over-committed. This can happen where there is overdevelopment of hydraulic infrastructure, usually for irrigation. Symptoms of physical water scarcity include environmental degradation and declining groundwater. Economic scarcity, meanwhile, is caused by a lack of investment in water or insufficient human capacity to satisfy the demand for water. Symptoms of economic water scarcity include a lack of infrastructure, with people often having to fetch water from rivers for domestic and agricultural uses. Some 2.8 billion people currently live in water-scarce areas.

Technical Challenges

Irrigation schemes involve solving numerous engineering and economic problems while minimizing negative environmental impact.

- Competition for surface water rights.

- Overdrafting (depletion) of underground aquifers.

- Ground subsidence (e.g. New Orleans, Louisiana)

- Underirrigation or irrigation giving only just enough water for the plant (e.g. in drip line irrigation) gives poor soil salinity control which leads to increased soil salinity with consequent buildup of toxic salts on soil surface in areas with high evaporation. This requires either leaching to remove these salts and a method of drainage to carry the salts away. When using drip lines, the leaching is best done regularly at certain intervals (with only a slight excess of water), so that the salt is flushed back under the plant's roots.

- Overirrigation because of poor distribution uniformity or management wastes water, chemicals, and may lead to water pollution.

- Deep drainage (from over-irrigation) may result in rising water tables which in some instances will lead to problems of irrigation salinity requiring watertable control by some form of subsurface land drainage.

- Irrigation with saline or high-sodium water may damage soil structure owing to the formation of alkaline soil

- Clogging of filters: It is mostly algae that clog filters, drip installations and nozzles. UV and ultrasonic method can be used for algae control in irrigation systems.

Theoretical Production Ecology

Theoretical production ecology tries to quantitatively study the growth of crops. The plant is treated as a kind of biological factory, which processes light, carbon dioxide, water, and nutrients into harvestable parts. Main parameters kept into consideration are temperature, sunlight, standing crop biomass, plant production distribution, nutrient and water supply.

Modelling

Modelling is essential in theoretical production ecology. Unit of modelling usually is the crop, the assembly of plants per standard surface unit. Analysis results for an individual plant are generalised to the standard surface, e.g. the Leaf Area Index is the projected surface area of all crop leaves above a unit area of ground.

Processes

The usual system of describing plant production divides the plant production process into at least five separate processes, which are influenced by several external parameters.

Two cycles of biochemical reactions constitute the basis of plant production, the light reaction and the dark reaction.

- In the light reaction, sunlight photons are absorbed by chloroplasts which split water into an electron, proton and oxygen radical which is recombined with another radical and released as molecular oxygen. The recombination of the electron with the proton yields the energy carriers NADH and ATP. The rate of this reaction often depends on sunlight intensity, leaf area index, leaf angle and amount of chloroplasts per leaf surface unit. The maximum theoretical gross production rate under optimum growth conditions is approximately 250 kg per hectare per day.

- The dark reaction or Calvin cycle ties atmospheric carbon dioxide and uses NADH and ATP to convert it into sucrose. The available NADH and ATP, as well as temperature and carbon dioxide levels determine the rate of this reaction. Together those two reactions are termed photosynthesis. The rate of photosynthesis is determined by the interaction of a number of factors including temperature, light intensity and carbon dioxide.

- The produced carbohydrates are transported to other plant parts, such as storage organs and converted into secondary products, such as amino acids, lipids, cellulose and other chemicals needed by the plant or used for respiration. Lipids, sugars, cellulose and starch can be produced without extra elements. The conversion of carbohydrates into amino acids and nucleic acids requires nitrogen, phosphorus and sulfur. Chlorophyll production requires magnesium, while several enzymes and coenzymes require trace elements. This means, nutrient supply influences this part of the production chain. Water supply is essential for transport, hence limits this too.

- The production centers, i.e. the leaves, are sources, the storage organs, growth tips or other destinations for the photosynthetic production are sinks. The lack of sinks can be a limiting factor for production too, as happens e.g. in apple orchards where insects or night frost have destroyed the blossoms and the produced assimilates cannot be converted into apples. Biennial and perennial plants employ the stored starch and fats in their storage organs to produce new leaves and shoots the next year.

- The amount of crop biomass and the relative distribution of biomass over leaves, stems, roots and storage organs determines the respiration rate. The amount of biomass in leaves determines the leaf area index, which is important in calculating the gross photosynthetic production.

- extensions to this basic model can include insect and pest damage, intercropping, climatical changes, etc.

Parameters

Important parameters in theoretical production models thus are:

Climate

- Temperature - The temperature determines the speed of respiration and the dark reaction. A high temperature combined with a low intensity of sunlight means a high loss by respiration. A low temperature combined with a high intensity of sunlight means that NADH and ATP heap up but cannot be converted into glucose because the dark reaction cannot process them swiftly enough.

- Light - Light, also called photosynthetic Active Radiation (PAR) is the energy source for green plant growth. PAR powers the light reaction, which provides ATP and NADPH for the conversion of carbon dioxide and water into carbohydrates and molecular oxygen. When temperature, moisture, carbon dioxide and nutrient levels are optimal, light intensity determines maximum production level.

- Carbon dioxide levels - Atmospheric carbon dioxide is the sole carbon source for plants. About half of all proteins in green leaves have the sole purpose of capturing carbon dioxide.

  Although CO_2 levels are constant under natural circumstances [on the contrary, CO2 concentration in the atmosphere has been increasing steadily for 200 years], CO_2 fertilization is common in greenhouses and is known to increase yields by on average 24% [a specific value, e.g., 24%, is meaningless without specification of the "low" and "high" CO2 levels being compared] .

  C_4 plants like maize and sorghum can achieve a higher yield at high solar radiation intensities, because they prevent the leaking of captured carbon dioxide due of the spatial separation of carbon dioxide capture and carbon dioxide use in the dark reaction. This means that their photorespiration is almost zero. This advantage is sometimes offset by a higher rate of maintenance respiration. In most models for natural crops, carbon dioxide levels are assumed to be constant.

Crop

- Standing crop biomass - Unlimited growth is an exponential process, which means that the amount of biomass determines the production. Because an increased biomass implies higher respiration per surface unit and a limited increase in intercepted light, crop growth is a sigmoid function of crop biomass.

- Plant production distribution - Usually only a fraction of the total plant biomass consists of useful products, e.g. the seeds in pulses and cereals, the tubers in potato and cassava, the leaves in sisal and spinach etc. The yield of usable plant portions will increase when the plant allocates more nutrients to this parts, e.g. the high-yielding varieties of wheat and rice allocate 40% of their biomass into wheat and rice grains, while the traditional varieties achieve only 20%, thus doubling the effective yield.

  Different plant organs have a different respiration rate, e.g. a young leaf has a much higher respiration rate than roots, storage tissues or stems do. There is a distinction between "growth respiration" and "maintenance respiration".

Sinks, such as developing fruits, need to be present. They are usually represented by a discrete switch, which is turned on after a certain condition, e.g. critical daylength has been met.

Care

- Water supply - Because plants use passive transport to transfer water and nutrients from their roots to the leaves, water supply is essential to growth, even so that water efficiency rates are known for different crops, e.g. 5000 for sugar cane, meaning that each kilogram of produced sugar requires up to 5000 liters of water.

- Nutrient supply - Nutrient supply has a twofold effect on plant growth. A limitation in nutrient supply will limit biomass production as per Liebig's Law of the Minimum. With some crops, several nutrients influence the distribution of plant products in the plants. A nitrogen gift is known to stimulate leaf growth and therefore can work adversely on the yield of crops which are accumulating photosynthesis products in storage organs, such as ripening cereals or fruit-bearing fruit trees.

Phases in Crop Growth

Theoretical production ecology assumes that the growth of common agricultural crops, such as cereals and tubers, usually consists of four (or five) phases:

- Germination - Agronomical research has indicated a temperature dependence of germination time (GT, in days). Each crop has a unique critical temperature (CT, dimension temperature) and temperature sum (dimensions temperature times time), which are related as follows.

$$GT = \frac{TS}{\sum_{k=1}^{N}(T - T_{crit})}$$

When a crop has a temperature sum of e.g. 150 °C·d and a critical temperature of 10 °C, it will germinate in 15 days when temperature is 20 °C, but in 10 days when temperature is 25 °C. When the temperature sum exceeds the threshold value, the germination process is complete.

- Initial spread - In this phase, the crop does not cover the field yet. The growth of the crop is linearly dependent on leaf area index, which in its turn is linearly dependent on crop biomass. As a result, crop growth in this phase is exponential.

- Total coverage of field - in this phase, growth is assumed to be linearly dependent on incident light and respiration rate, as nearly 100% of all incident light is intercepted. Typically, the Leaf Area Index (LAI) is above two to three in this phase. This phase of vegetative growth ends when the plant gets a certain environmental or internal signal and starts generative growth (as in cereals and pulses) or the storage phase (as in tubers).

- Allocation to storage organs - in this phase, up to 100% of all production is directed to the storage organs. Generally, the leaves are still intact and as a result, gross primary production stays the same. Prolonging this phase, e.g. by careful fertilization, water and pest management results directly in a higher harvest.

- Ripening - in this phase, leaves and other production structures slowly die off. Their carbohydrates and proteins are transported to the storage organs. As a result, the LAI and, hence, the primary production decreases.

Existing Plant Production Models

Plant production models exist in varying levels of scope (cell, physiological, individual plant, crop, geographical region, global) and of generality: the model can be crop-specific or be more generally applicable. In this section the emphasis will be on crop-level based models as the crop is the main area of interest from an agronomical point of view.

As of 2005, several crop production models are in use. The crop growth model SUCROS has been developed during more than 20 years and is based on earlier models. Its latest revision known dates from 1997. The IRRI and Wageningen University more recently developed the rice growth model ORYZA2000. This model is used for modeling rice growth. Both crop growth models are open source. Other more crop-specific plant growth models exist as well.

SUCROS

SUCROS is programmed in the Fortran computer programming language. The model can and has been applied to a variety of weather regimes and crops. Because the source code of Sucros is open source, the model is open to modifications of users with FORTRAN programming experience. The official maintained version of SUCROS comes into two flavours: SUCROS I, which has non-inhibited unlimited crop growth (which means that only solar radiation and temperature determine growth) and SUCROS II, in which crop growth is limited only by water shortage.

ORYZA2000

The ORYZA2000 rice growth model has been developed at the IRRI in cooperation with Wageningen University. This model, too, is programmed in FORTRAN. The scope of this model is limited to rice, which is the main food crop for Asia.

Other Models

The United States Department of Agriculture has sponsored a number of applicable crop growth models for various major US crops, such as cotton, soy bean, wheat and rice. Other widely used models are the precursor of SUCROS (SWATR), CERES, several incarnations of PLANTGRO, SUBSTOR, the FAO-sponsored CROPWAT, AGWATER and the erosion-specific model EPIC. , cropping system CropSyst

A less mechanistic growth and competition model, called the Conductance Model, has been developed, mainly at Warwick-HRI, Wellesbourne, UK. This model simulates light interception and growth of individual plants based on the lateral expansion of their crown zone areas. Competition between plants is simulated by a set algorithms related to competition for space and resultant light intercept as the canopy closes. Some versions of the model assume overtopping of some species by others. Although the model cannot take account of water or mineral nutrients, it can simulate individual plant growth, variability in growth within plant communities and inter-species competition. This model was written in Matlab.

Leaf Area Index

Leaf area index (LAI) is a dimensionless quantity that characterizes plant canopies. It is defined as the one-sided green leaf area per unit ground surface area (*LAI = leaf area / ground area, m² / m²*) in broadleaf canopies. In conifers, three definitions for LAI have been used:

- Half of the total needle surface area per unit ground surface area

- Projected (or one-sided, in accordance the definition for broadleaf canopies) needle area per unit ground area

- Total needle surface area per unit ground area

LAI ranges from 0 (bare ground) to over 10 (dense conifer forests).

Interpretation and Application of LAI

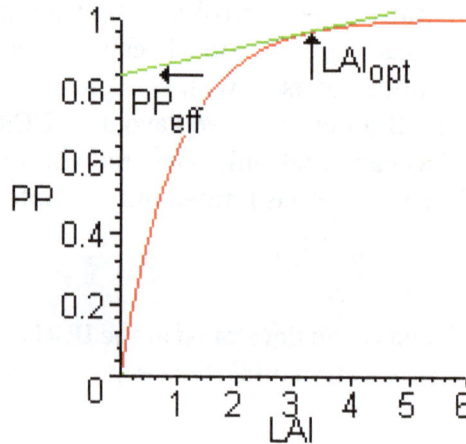

Effective PP, obtained by subtracting respiration

LAI is used to predict photosynthetic primary production, evapotranspiration and as a reference tool for crop growth. As such, LAI plays an essential role in theoretical production ecology. An inverse exponential relation between LAI and light interception, which is linearly proportional to the primary production rate, has been established:

$$P = P_{max} \left(1 - e^{-c \cdot LAI}\right)$$

where P_{max} designates the maximum primary production and c designates a crop-specific growth coefficient. This inverse exponential function is called the primary production function.

Determining LAI

LAI can be determined directly by taking a statistically significant sample of foliage from a plant canopy, measuring the leaf area per sample plot and dividing it by the plot land surface area. Indirect methods measure canopy geometry or light extinction and relate it to LAI.

Direct Methods

Direct methods can be easily applied on deciduous species by collecting leaves during leaf fall in traps of certain area distributed below the canopy. The area of the collected leaves can be measured using a leaf area meter or an image scanner and image analysis software. The measured leaf area can then be divided by the area of the traps to obtain LAI. Alternatively, leaf area can be measured on a sub-sample of the collected leaves and linked to the leaf dry mass (e.g. via Specific Leaf Area, SLA cm²/g). That way it is not necessary to measure the area of all leaves one by one, but weigh the collected leaves after drying (at 60–80 °C for 48 h). Leaf dry mass multiplied by the specific leaf area is converted into leaf area. Direct methods in evergreen species are necessarily destructive. However, they are widely used in crops and pastures by harvesting the vegetation and measuring leaf area within a certain ground surface area. It is very difficult (and also unethical) to apply such destructive techniques in natural ecosystems, particularly in forests of evergreen tree species. Foresters have developed techniques that determine leaf area in evergreen forests through allometric relationships. Due to the difficulties and the limitations of the direct methods for estimating LAI, they are mostly used as reference for indirect methods that are easier and faster to apply.

Indirect Methods

A hemispherical photograph of forest canopy. The ratio of the area of canopy to sky is used to approximate LAI.

Indirect methods of estimating LAI *in situ* can be divided in two categories:

1. indirect contact LAI measurements such as plumb lines and inclined point quadrats

2. indirect non-contact measurements

Due to the subjectivity and labor involved with the first method, indirect non-contact measurements are typically preferred. Non-contact LAI tools, such as hemispherical photography,

Hemiview Plant Canopy Analyser from Delta-T Devices, the CI-110 Plant Canopy Analyzer from CID Bio-Science, LAI-2200 Plant Canopy Analyzer from LI-COR Biosciences and the LP-80 LAI ceptometer from Decagon Devices, measure LAI in a non-destructive way. Hemispherical photography methods estimate LAI and other canopy structure attributes from analyzing upward-looking fisheye photographs taken beneath the plant canopy. The LAI-2200 calculates LAI and other canopy structure attributes from solar radiation measurements made with a wide-angle optical sensor. Measurements made above and below the canopy are used to determine canopy light interception at five angles, from which LAI is computed using a model of radiative transfer in vegetative canopies. The LP-80 calculates LAI by means of measuring the difference between light levels above the canopy and at ground level, and factoring in the leaf angle distribution, solar zenith angle, and plant extinction coefficient. Such indirect methods, where LAI is calculated based upon observations of other variables (canopy geometry, light interception, leaf length and width, etc.) are generally faster, amenable to automation, and thereby allow for a larger number of spatial samples to be obtained. For reasons of convenience when compared to the direct (destructive) methods, these tools are becoming more and more important.

Disadvantages of Methods

The disadvantage of the direct method is that it is destructive, time consuming and expensive, especially if the study area is very large.

The disadvantage of the indirect method is that in some cases it can underestimate the value of LAI in very dense canopies, as it does not account for leaves that lie on each other, and essentially act as one leaf according to the theoretical LAI models. Ignorance of non-randomness within canopies may cause underestimation of LAI up to 25%, introducing path length distribution in the indirect method can improve the measuring accuracy of LAI.

Animal Feed

A photo of a feedlot in Texas, USA, where cattle are "finished" (fattened on grains) prior to slaughter.

Animal feed is food given to domestic animals in the course of animal husbandry. There are two basic types, *fodder* and *forage*. Used alone, the word "feed" more often refers to fodder.

Fodder

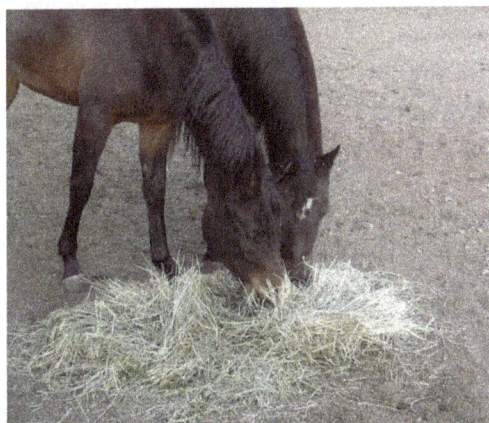

Equine nutritionists recommend that 50% or more of a horse's diet by weight should be forages, such as hay

"Fodder" refers particularly to foods or forages given to the animals (including plants cut and carried to them), rather than that which they forage for themselves. It includes hay, straw, silage, compressed and pelleted feeds, oils and mixed rations, and sprouted grains and legumes. Feed grains are the most important source of animal feed globally. The amount of grain used to produce the same unit of meat varies substantially. According to an estimate reported by the BBC in 2008, "Cows and sheep need 8kg of grain for every 1kg of meat they produce, pigs about 4kg. The most efficient poultry units need a mere 1.6kg of feed to produce 1kg of chicken." Farmed fish can also be fed on grain, and use even less than poultry. The two most important feed grains are maize and soyabean, and the United States is by far the largest exporter of both, averaging about half of the global maize trade and 40% of the global soya trade in the years leading up the 2012 drought. Other feed grains include wheat, oats, barley, and rice, among many others.

Traditional sources of animal feed include household food scraps and the byproducts of food processing industries such as milling and brewing. Material remaining from milling oil crops like peanuts, soy, and corn are important sources of fodder. Scraps fed to pigs are called slop, and those fed to chicken are called chicken scratch. Brewer's spent grain is a byproduct of beer making that is widely used as animal feed.

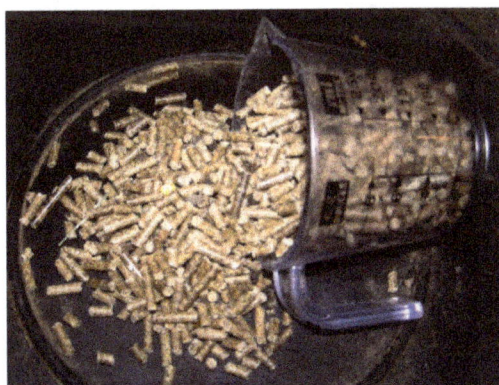

A pelleted ration designed for horses

Compound feed is fodder that is blended from various raw materials and additives. These blends are formulated according to the specific requirements of the target animal. They are manufactured by feed compounders as *meal type*, *pellets* or *crumbles*. The main ingredients used in commercially prepared feed are the feed grains, which include corn, soybeans, sorghum, oats, and barley.

Compound feed may also include premixes, which may also be sold separately. Premixes are composed of microingredients such as vitamins, minerals, chemical preservatives, antibiotics, fermentation products, and other essential ingredients that are purchased from premix companies, usually in sacked form, for blending into commercial rations. Because of the availability of these products, a farmer who uses his own grain can formulate his own rations and be assured his animals are getting the recommended levels of minerals and vitamins.

According to the American Feed Industry Association, as much as $20 billion worth of feed ingredients are purchased each year. These products range from grain mixes to orange rinds to beet pulps. The feed industry is one of the most competitive businesses in the agricultural sector, and is by far the largest purchaser of U.S. corn, feed grains, and soybean meal. Tens of thousands of farmers with feed mills on their own farms are able to compete with huge conglomerates with national distribution. Feed crops generated $23.2 billion in cash receipts on U.S. farms in 2001. At the same time, farmers spent a total of $24.5 billion on feed that year.

In 2011, around 734.5 million tons of feed were produced annually around the world.

History

The beginning of industrial-scale production of animal feeds can be traced back to the late 19th century, around the time advances in human and animal nutrition were able to identify the benefits of a balanced diet, and the importance of the role processing of certain raw materials played. Corn gluten feed was first manufactured in 1882, while leading world feed producer Purina Feeds was established in 1894 by William Hollington Danforth. Cargill, which was mainly dealing in grains from its beginnings in 1865, started to deal in feed at about 1884.

The feed industry expanded rapidly in the first quarter of the 20th century, with Purina expanding its operations into Canada, and opened its first feed mill in 1927 (which is still in operation). In 1928, the feed industry was revolutionized by the introduction of the first pelleted feeds - Purina Checkers.

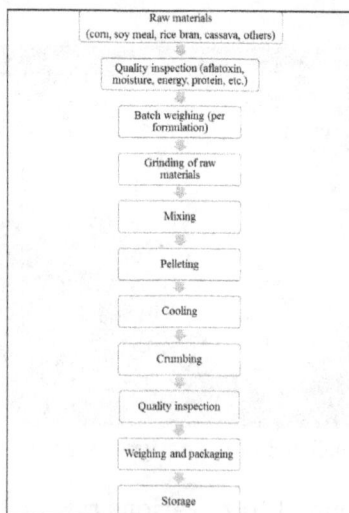

Workflow for general feed manufacturing process

Cattle eating a total mixed ration

Manufacturers

The job of the feed manufacturer is to buy the commodities and blend them in the feed mill according to the specifications outlined by the animal nutritionist. There is little room for error because, if the ration is not apportioned correctly, lowered animal production and diminished outward appearance can occur.

Asia

One of the largest Asian feed producers is Charoen Pokphand (the CP Group), a Thai company producing 18 million tonnes of compound feed at various locations across East Asia.

Europe

The merge of the Hamburg-based traditional commodity trade firm, Cremer, and the Düsseldorf based Deuka (Deutsche Kraftfutterwerke), led to one of the largest feed companies in Europe. The new Cremer Group produces around 3.5 million tons. BOCM Pauls in the UK produces around the same amount if not more.

United States

Feed production facility in Oneonta, New York

Leading U.S. companies involved in prepared feeds production in the early first decade of the 21st century included ConAgra Inc., an Omaha, Nebraska-based firm; and Cargill, Incorporated, a diversified company that was the nation's top exporter of grain. In 1998, Ralston Purina Company, based in St. Louis, Missouri, formed Agribrands International, Inc. to control its international animal feed and agricultural products division. Agribrands produced feed and other products for livestock in markets outside of the United States, and had about 75 facilities operating in 16 countries. In 2001, it was acquired by Cargill.

Other significant industry players included Conti Group Companies, Inc., the world's leading cattle feeder; CHS, Inc. (previously known as Cenex Harvest States Cooperative), which was primarily involved in grain trading; and Farmland Industries, Inc., the leading agricultural cooperative in the United States. Farmland was a worldwide exporter of products, such as grain. In May 2002, the firm declared bankruptcy, and in the following year, Smithfield Foods acquired most of Farmland's assets.

Forage

In this photo a herdsman from the Masaai people watches as his cattle graze in the Ngorongoro crater, Tanzania.

"Forage" is plant material (mainly plant leaves and stems) eaten by grazing livestock. Historically, the term *forage* has meant only plants eaten by the animals directly as pasture, crop residue, or immature cereal crops, but it is also used more loosely to include similar plants cut for fodder and carried to the animals, especially as hay or silage.

Manufacture

Nutrition

In agriculture today, the nutritional needs of farm animals are well understood and may be satisfied through natural forage and fodder alone, or augmented by direct supplementation of nutrients in concentrated, controlled form. The nutritional quality of feed is influenced not only by the nutrient content, but also by many other factors such as feed presentation, hygiene, digestibility, and effect on intestinal health.

Feed additives provide a mechanism through which these nutrient deficiencies can be resolved effect the rate of growth of such animals and also their health and well-being. Even with all the benefits of higher quality feed, most of a farm animal's diet still consists of grain-based ingredients because of the higher costs of quality feed.

By Animal

- Bird food

- Cat food

- Cattle feeding

- Dog food

- Equine nutrition

- Pet food

- Pig farming

- Poultry feed

- Sheep husbandry

References

- Frenken, K. (2005). Irrigation in Africa in figures – AQUASTAT Survey – 2005 (PDF). Food and Agriculture Organization of the United Nations. ISBN 92-5-105414-2. Retrieved 2007-03-14.

- Drainage Manual: A Guide to Integrating Plant, Soil, and Water Relationships for Drainage of Irrigated Lands. Interior Dept., Bureau of Reclamation. 1993. ISBN 0-16-061623-9.

- Fox Business News: "Machinery takes the place of migrants as Maine's blueberry harvest booms" September 06, 2015.

- Las Cruces Sun: "Experts: Machines could reverse declining New Mexico green chile acreage" By Diana Alba Soular July 25, 2015.

- Fresno Bee: "New raisin grape holds promise for central San Joaquin Valley growers" By Robert Rodriguez September 19, 2015.

- "No Hands Touch the Land: Automating California Farms" (PDF). California Agrarian Action Project: 20–28. July 1977. Retrieved 2015-04-25.

- "A new report says we're draining our aquifers faster than ever". High Country News. 2013-06-22. Retrieved 2014-02-11.

7

Soil Nutrition and Soil Sampling

The ability of soil to nurture the growth of a plant is known as soil fertility. Soil conversation is the prevention of soil loss from erosion. The methods of soil conservation include crop rotation, cover crop, tillage and windbreak. In order to completely understand soil nutrition and soil sampling, it is necessary to understand the processes related to it. The following section elucidates the processes associated with soil nutrition.

Soil Fertility

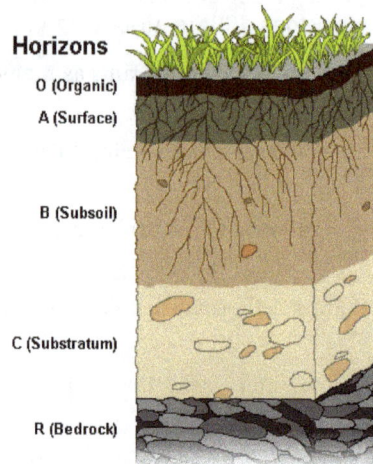

Horizons
O (Organic)
A (Surface)
B (Subsoil)
C (Substratum)
R (Bedrock)

Soil scientists use the capital letters O, A, B, C, and E to identify the master horizons, and lowercase letters for distinctions of these horizons. Most soils have three major horizons—the surface horizon (A), the subsoil (B), and the substratum (C). Some soils have an organic horizon (O) on the surface, but this horizon can also be buried. The master horizon, E, is used for subsurface horizons that have a significant loss of minerals (eluviation). Hard bedrock, which is not soil, uses the letter R.

Soil fertility refers to the ability of a soil to sustain plant growth, i.e. to provide plant habitat and result in sustained and consistent yields of high quality. A fertile soil has the following properties:

- The ability to supply essential plant nutrients and soil water in adequate amounts and proportions for plant growth and reproduction; and

- The absence of toxic substances which may inhibit plant growth.

The following properties contribute to soil fertility in most situations:

- Sufficient soil depth for adequate root growth and water retention;

- Good internal drainage, allowing sufficient aeration for optimal root growth (although some plants, such as rice, tolerate waterlogging);

- Topsoil with sufficient soil organic matter for healthy soil structure and soil moisture retention;

- Soil pH in the range 5.5 to 7.0 (suitable for most plants but some prefer or tolerate more acid or alkaline conditions);

- Adequate concentrations of essential plant nutrients in plant-available forms;

- Presence of a range of microorganisms that support plant growth.

In lands used for agriculture and other human activities, maintenance of soil fertility typically requires the use of soil conservation practices. This is because soil erosion and other forms of soil degradation generally result in a decline in quality with respect to one or more of the aspects indicated above.

Soil Fertilization

Bioavailable phosphorus is the element in soil that is most often lacking. Nitrogen and potassium are also needed in substantial amounts. For this reason these three elements are always identified on a commercial fertilizer analysis. For example, a 10-10-15 fertilizer has 10 percent nitrogen, 10 percent (P_2O_5) available phosphorus and 15 percent (K_2O) water-soluble potassium. Sulfur is the fourth element that may be identified in a commercial analysis—e.g. 21-0-0-24 which would contain 21% nitrogen and 24% sulfate.

Inorganic fertilizers are generally less expensive and have higher concentrations of nutrients than organic fertilizers. Also, since nitrogen, phosphorus and potassium generally must be in the inorganic forms to be taken up by plants, inorganic fertilizers are generally immediately bioavailable to plants without modification. However, some have criticized the use of inorganic fertilizers, claiming that the water-soluble nitrogen doesn't provide for the long-term needs of the plant and creates water pollution. Slow-release fertilizers may reduce leaching loss of nutrients and may make the nutrients that they provide available over a longer period of time.

Soil fertility is a complex process that involves the constant cycling of nutrients between organic and inorganic forms. As plant material and animal wastes are decomposed by micro-organisms, they release inorganic nutrients to the soil solution, a process referred to as mineralization. Those nutrients may then undergo further transformations which may be aided or enabled by soil micro-organisms. Like plants, many micro-organisms require or preferentially use inorganic forms of nitrogen, phosphorus or potassium and will compete with plants for these nutrients, tying up the nutrients in microbial biomass, a process often called immobilization. The balance between immobilization and mineralization processes depends on the balance and availability of major nutrients and organic carbon to soil microorganisms. Natural processes such as lightning strikes may fix atmospheric nitrogen by converting it to (NO_2). Denitrification may occur under anaerobic conditions (flooding) in the presence of denitrifying bacteria. The cations, primarily phosphate and potash, as well as many micronutrients are held in relatively strong bonds with the negatively charged portions of the soil in a process known as cation-exchange capacity.

In 2008 the cost of phosphorus as fertilizer more than doubled, while the price of rock phosphate as base commodity rose eight-fold. Recently the term peak phosphorus has been coined, due to the limited occurrence of rock phosphate in the world.

Light and CO2 Limitations

Photosynthesis is the process whereby plants use light energy to drive chemical reactions which convert CO_2 into sugars. As such, all plants require access to both light and carbon dioxide to produce energy, grow and reproduce.

While typically limited by nitrogen, phosphorus and potassium, low levels of carbon dioxide can also act as a limiting factor on plant growth. Peer-reviewed and published scientific studies have shown that increasing CO_2 is highly effective at promoting plant growth up to levels over 300 ppm. Further increases in CO_2 can, to a very small degree, continue to increase net photosynthetic output.

Since higher levels of CO_2 have only a minimal impact on photosynthetic output at present levels (presently around 400 ppm and increasing), we should not consider plant growth to be limited by carbon dioxide. Other biochemical limitations, such as soil organic content, nitrogen in the soil, phosphorus and potassium, are far more often in short supply. As such, neither commercial nor scientific communities look to air fertilization as an effective or economic method of increasing production in agriculture or natural ecosystems. Furthermore, since microbial decomposition occurs faster under warmer temperatures, higher levels of CO_2 (which is one of the causes of unusually fast climate change) should be expected to increase the rate at which nutrients are leached out of soils and may have a negative impact on soil fertility.

Soil Depletion

Soil depletion occurs when the components which contribute to fertility are removed and not replaced, and the conditions which support soil's fertility are not maintained. This leads to poor crop yields. In agriculture, depletion can be due to excessively intense cultivation and inadequate soil management.

Soil fertility can be severely challenged when land use changes rapidly. For example, in Colonial New England, colonists made a number of decisions that depleted the soils, including: allowing herd animals to wander freely, not replenishing soils with manure, and a sequence of events that led to erosion. William Cronon wrote that "...the long-term effect was to put those soils in jeopardy. The removal of the forest, the increase in destructive floods, the soil compaction and close-cropping wrought by grazing animals, plowing--all served to increase erosion."

One of the most widespread occurrences of soil depletion as of 2008 is in tropical zones where nutrient content of soils is low. The combined effects of growing population densities, large-scale industrial logging, slash-and-burn agriculture and ranching, and other factors, have in some places depleted soils through rapid and almost total nutrient removal.

Topsoil depletion occurs when the nutrient-rich organic topsoil, which takes hundreds to thousands of years to build up under natural conditions, is eroded or depleted of its original organic material. Historically, many past civilizations' collapses can be attributed to the depletion of the topsoil. Since the beginning of agricultural production in the Great Plains of North America in the 1880s, about one-half of its topsoil has disappeared.

Depletion may occur through a variety of other effects, including overtillage (which damages soil structure), underuse of nutrient inputs which leads to mining of the soil nutrient bank, and salinization of soil.

Irrigation Water Effects

The quality of irrigation water is very important to maintain soil fertility and tilth, and for using more soil depth by the plants. When soil is irrigated with high alkaline water, unwanted sodium salts build up in the soil which would make soil draining capacity very poor. So plant roots can not penetrate deep in to the soil for optimum growth in Alkali soils. When soil is irrigated with low pH / acidic water, the useful salts (Ca, Mg, K, P, S, etc.) are removed by draining water from the acidic soil and in addition unwanted aluminium and manganese salts to the plants are dissolved from the soil impeding plant growth. When soil is irrigated with high salinity water or sufficient water is not draining out from the irrigated soil, the soil would convert in to saline soil or lose its fertility. Saline water enhance the turgor pressure or osmotic pressure requirement which impedes the off take of water and nutrients by the plant roots.

Top soil loss takes place in alkali soils due to erosion by rain water surface flows or drainage as they form colloids (fine mud) in contact with water. Plants absorb water-soluble inorganic salts only from the soil for their growth. Soil as such does not lose fertility just by growing crops but it lose its fertility due to accumulation of unwanted and depletion of wanted inorganic salts from the soil by improper irrigation and acid rain water (quantity and quality of water). The fertility of many soils which are not suitable for plant growth can be enhanced many times gradually by providing adequate irrigation water of suitable quality and good drainage from the soil.

Global Distribution

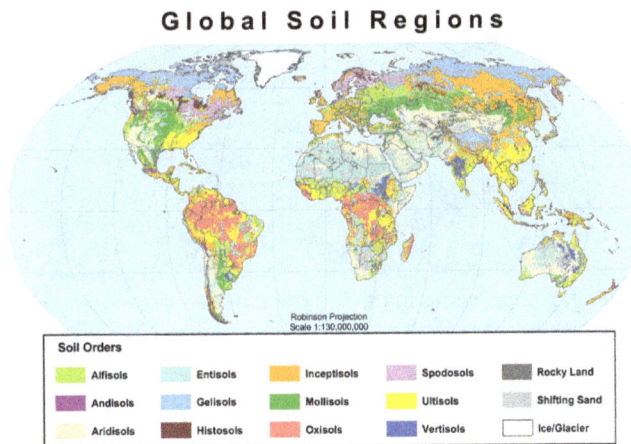

Global distribution of soil types of the USDA soil taxonomy system. Mollisols, shown here in dark green, are a good (though not the only) indicator of high soil fertility. They coincide to a large extent with the world's major grain producing areas like the North American Prairie States, the Pampa and Gran Chaco of South America and the Ukraine-to-Central Asia Black Earth belt.

Soil Conservation

Soil conservation is the prevention of soil loss from erosion or reduced fertility caused by over usage, acidification, salinization or other chemical soil contamination.

Slash-and-burn and other unsustainable methods of subsistence farming are practiced in some lesser developed areas. A sequel to the deforestation is typically large scale erosion, loss of soil nutrients and sometimes total desertification.

Erosion barriers on disturbed slope, Marin County, California

Techniques for improved soil conservation include crop rotation, cover crops, conservation tillage and planted windbreaks and affect both erosion and fertility. When plants, especially trees, die, they decay and become part of the soil. Code 330 defines standard methods recommended by the US Natural Resources Conservation Service.

Contour plowing, Pennsylvania 1938. The rows formed slow water run-off during rainstorms to prevent soil erosion and allows the water time to infiltrate into the soil.

Farmers have practiced soil conservation for millennia. Conservation practices fall in multiple categories:

Contour Plowing

Contour plowing orients crop furrows following the contour lines of the farmed area. Furrows move left and right to maintain a constant altitude, which reduces runoff. Contour plowing was practiced by the ancient Phoenicians, and is effective for slopes between two and ten percent. Contour plowing can increase crop yields from 10 to 50 percent, partially as a result from greater soil retention.

Terracing or Terrace Farming

Terracing is the practice of creating nearly level areas in a hillside area. The terraces form a series of steps, each at a higher level than the previous. Terraces are protected from erosion by other soil barriers.

Terraced farming is more common on small farms and in underdeveloped countries, since mechanized equipment is difficult to deploy in this setting. It protects the soil from its erosion. It is one of the way by which soil erosion can be stopped.

Keyline Design

Keyline design is an enhancement of contour farming, where the total watershed properties are taken into account in forming the contour lines.

Perimeter Runoff Control

Tree, shrubs and ground-cover are effective perimeter treatment for soil erosion prevention, by impeding surface flows. A special form of this perimeter or inter-row treatment is the use of a "grass way" that both channels and dissipates runoff through surface friction, impeding surface runoff and encouraging infiltration of the slowed surface water.

Windbreaks

Windbreaks are sufficiently dense rows of trees at the windward exposure of an agricultural field subject to wind erosion. Evergreen species provide year-round protection; however, as long as foliage is present in the seasons of bare soil surfaces, the effect of deciduous trees may be adequate.

Cover Crops/Crop Rotation

Cover crops such as legumes, white turnip, radishes and other species are rotated with cash crops to blanket the soil year-round and act as green manure that replenishes nitrogen and other critical nutrients. Cover crops also help suppress weeds.

Soil-conservation Farming

Soil-conservation farming involves no-till farming, "green manures" and other soil-enhancing practices. Such farming methods attempt to mimic the biology of virgin land. They can revive damaged soil, minimize erosion, encourage plant growth, eliminate the use of nitrogen fertilizer or fungicide, produce above-average yields and protect crops during droughts or flooding. The result is less labor and lower costs that increase farmers' profits. No-till farming and cover crops act as sinks for nitrogen and other nutrients. This increases the amount of soil organic matter.

Repeated plowing/tilling degrades soil, killing its beneficial fungi and earthworms. Once damaged, soil may take multiple seasons to fully recover, even in optimal circumstances.

Critics argue that no-till and related methods are impractical and too expensive for many growers, partly because it requires new equipment. They cite advantages for conventional tilling depending on the geography, crops and soil conditions. Some farmers claimed that no-till complicates weed control, delays planting and that post-harvest residues, especially for corn, are hard to manage.

Salinity Management

Salt deposits on the former bed of the Aral Sea

Salinity in soil is caused by irrigating with salty water. Water then evaporates from the soil leaving the salt behind. Salt breaks down the soil structure, causing infertility and reduced growth.

The ions responsible for salination are: sodium (Na^+), potassium (K^+), calcium (Ca^{2+}), magnesium (Mg^{2+}) and chlorine (Cl^-). Salinity is estimated to affect about one third of the earth's arable land. Soil salinity adversely affects crop metabolism and erosion usually follows.

Salinity occurs on drylands from overirrigation and in areas with shallow saline water tables. Over-irrigation deposits salts in upper soil layers as a byproduct of soil infiltration; irrigation merely increases the rate of salt deposition. The best-known case of shallow saline water table capillary action occurred in Egypt after the 1970 construction of the Aswan Dam. The change in the groundwater level led to high salt concentrations in the water table. The continuous high level of the water table led to soil salination.

Use of humic acids may prevent excess salination, especially given excessive irrigation. Humic acids can fix both anions and cations and eliminate them from root zones. Planting species that can tolerate saline conditions to produce surface cover can preserve soil salinity can be reduced. Salt-tolerant plants include saltbush, a plant found in much of North America and in the Mediterranean regions of Europe.

Soil Organisms

Yellow fungus, a mushroom that assists in organic decay.

When worms excrete egesta in the form of casts, a balanced selection of minerals and plant nutrients is made into a form accessible for root uptake. Earthworm casts are five times richer in available nitrogen, seven times richer in available phosphates and eleven times richer in available potash than the surrounding upper 150 millimetres (5.9 in) of soil. The weight of casts produced may be greater than 4.5 kg per worm per year. By burrowing, the earthworm improves soil porosity, creating channels that enhance the processes of aeration and drainage.

Other important soil organisms include nematodes, mycorrizha and bacteria.

Degraded soil requires synthetic fertilizer to produce high yields. Lacking structure increases erosion and carries nitrogen and other pollutants into rivers and streams.

Each one percent increase in soil organic matter helps soil hold 20,000 gallons more water per acre.

Mineralization

To allow plants full realization of their phytonutrient potential, active mineralization of the soil is sometimes undertaken. This can involve adding crushed rock or chemical soil supplements. In either case the purpose is to combat mineral depletion. A broad range of minerals can be used, including common substances such as phosphorus and more exotic substances such as zinc and selenium. Extensive research examines the phase transitions of minerals in soil with aqueous contact.

Flooding can bring significant sediment to an alluvial plain. While this effect may not be desirable if floods endanger life or if the sediment originates from productive land, this process of addition to a floodplain is a natural process that can rejuvenate soil chemistry through mineralization and macronutrient addition.

Methods of Soil Conservation

Crop Rotation

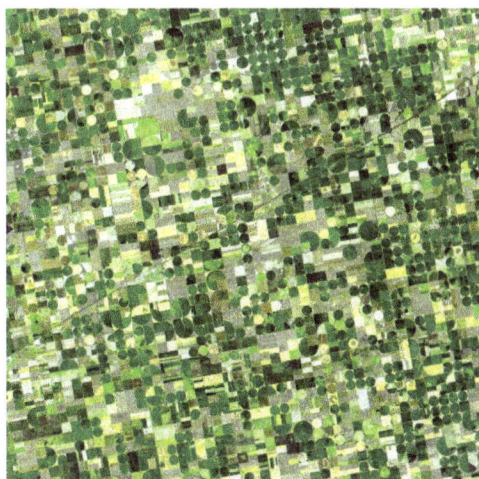

Satellite image of circular crop fields in Kansas in late June 2001. Healthy, growing crops are green. Corn would be growing into leafy stalks by then. Sorghum, which resembles corn, grows more slowly and would be much smaller and therefore, (possibly) paler. Wheat is a brilliant yellow as harvest occurs in June. Fields of brown have been recently harvested and plowed under or lie fallow for the year.

Effects of crop rotation and monoculture at the Swojec Experimental Farm, Wroclaw University of Environmental and Life Sciences. In the front field, the "Norfolk" crop rotation sequence (potatoes, oats, peas, rye) is being applied; in the back field, rye has been grown for 45 years in a row.

Crop rotation is the practice of growing a series of dissimilar or different types of crops in the same area in sequenced seasons. It is done so that the soil of farms is not used to only one type of nutrient. It helps in reducing soil erosion and increases soil fertility and crop yield.

Growing the same crop in the same place for many years in a row disproportionately depletes the soil of certain nutrients. With rotation, a crop that leaches the soil of one kind of nutrient is followed during the next growing season by a dissimilar crop that returns that nutrient to the soil or draws a different ratio of nutrients. In addition, crop rotation mitigates the buildup of pathogens and pests that often occurs when one species is continuously cropped, and can also improve soil structure and fertility by increasing biomass from varied root structures.

Crop rotation is used in both conventional and organic farming systems.

History

Agriculturalists have long recognized that suitable rotations – such as planting spring crops for livestock in place of grains for human consumption – make it possible to restore or to maintain a productive soil. Middle Eastern farmers practised crop rotation in 6000 BC without understanding the chemistry, alternately planting legumes and cereals. In the Bible, chapter 25 of the Book of Leviticus instructs the Israelites to observe a "Sabbath of the Land". Every seventh year they would not till, prune or even control insects. The Roman writer, Cato the Elder (234 – 149 BC), recommended that farmers "save carefully goat, sheep, cattle, and all other dung". From the times of Charlemagne (died 814), farmers in Europe transitioned from a two-field crop rotation to a three-field crop rotation. Under a two-field rotation, half the land was planted in a year, while the other half lay fallow. Then, in the next year, the two fields were reversed.

From the end of the Middle Ages until the 20th century, Europe's farmers practised three-field rotation, dividing available lands into three parts. One section was planted in the autumn with rye or winter wheat, followed by spring oats or barley; the second section grew crops such as peas, lentils, or beans; and the third field was left fallow. The three fields were rotated in this manner so that every three years, a field would rest and be fallow. Under the two-field system, if one has a total of 600 acres (2.4 km²) of fertile land, one would only plant 300 acres. Under the new three-field rotation system, one would plant (and therefore harvest) 400 acres. But the additional crops had a more significant effect than mere quantitative productivity. Since the spring crops were mostly legumes, they increased the overall nutrition of the people of Northern Europe.

Farmers in the region of Waasland (in present-day northern Belgium) pioneered a four-field rotation in the early 16th century, and the British agriculturist Charles Townshend (1674-1738) popularised this system in the 18th century. The sequence of four crops (wheat, turnips, barley and clover), included a fodder crop and a grazing crop, allowing livestock to be bred year-round. The four-field crop rotation became a key development in the British Agricultural Revolution.

George Washington Carver (1860s - 1943) studied crop-rotation methods in the United States, teaching southern farmers to rotate soil-depleting crops like cotton with soil-enriching crops like peanuts and peas.

In the Green Revolution of the mid-20th century the traditional practice of crop rotation gave way in some parts of the world to the practice of supplementing the chemical inputs to the soil through topdressing with fertilizers, adding (for example) ammonium nitrate or urea and restoring soil pH with lime. Such practices aimed to increase yields, to prepare soil for specialist crops, and to reduce waste and inefficiency by simplifying planting and harvesting.

Crop Choice

A preliminary assessment of crop interrelationships can be found in how each crop: (1) contributes to soil organic matter (SOM) content, (2) provides for pest management, (3) manages deficient or excess nutrients, and (4) how it contributes to or controls for soil erosion.

Crop choice is often related to the goal the farmer is looking to achieve with the rotation, which could be weed management, increasing available nitrogen in the soil, controlling for erosion, or increasing soil structure and biomass, to name a few. When discussing crop rotations, crops are classified in different ways depending on what quality is being assessed: by family, by nutrient needs/benefits, and/or by profitability (i.e. cash crop versus cover crop). For example, giving adequate attention to plant family is essential to mitigating pests and pathogens. However, many farmers have success managing rotations by planning sequencing and cover crops around desirable cash crops. The following is a simplified classification based on crop quality and purpose.

Row Crops

Many crops which are critical for the market, like vegetables, are row crops (that is, grown in tight rows). While often the most profitable for farmers, these crops are more taxing on the soil Row crops typically have low biomass and shallow roots: this means the plant contributes low residue to the surrounding soil and has limited effects on structure. With much of the soil around the plant is exposed to disruption by rainfall and traffic, fields with row crops experience faster break down of organic matter by microbes, leaving fewer nutrients for future plants.

In short, while these crops may be profitable for the farm, they are nutrient depleting. Crop rotation practices exist to strike a balance between short-term profitability and long-term productivity.

Legumes

A great advantage of crop rotation comes from the interrelationship of nitrogen fixing-crops with nitrogen demanding crops. Legumes, like alfalfa and clover, collect available nitrogen from the soil in nodules on their root structure. When the plant is harvested, the biomass of uncollected roots breaks down, making the stored nitrogen available to future crops. Legumes are also a valued green manure: a crop that collects nutrients and fixes them at soil depths accessible to future crops.

In addition, legumes have heavy tap roots that burrow deep into the ground, lifting soil for better tilth and absorption of water.

Grasses and Cereals

Cereal and grasses are frequent cover crops because of the many advantages they supply to soil quality and structure. The dense and far-reaching root systems give ample structure to surrounding soil and provide significant biomass for soil organic matter.

Grasses and cereals are key in weed management as they compete with undesired plants for soil space and nutrients.

Green Manure

Green manure is a crop that is mixed into the soil. Both nitrogen-fixing legumes and nutrient scavengers, like grasses, can be used as green manure. Green manure of legumes is an excellent source of nitrogen, especially for organic systems, however, legume biomass doesn't contribute to lasting soil organic matter like grasses do.

Planning a Rotation

There are numerous factors that must be taken into consideration when planning a crop rotation. Planning an effective rotation requires weighing fixed and fluctuating production circumstances, including, but not limited to: market, farm size, labor supply, climate, soil type, growing practices, etc. Moreover, a crop rotation must consider in what condition one crop will leave the soil for the succeeding crop and how one crop can be seeded with another crop. For example, a nitrogen-fixing crop, like a legume, should always proceed a nitrogen depleting one; similarly, a low residue crop (i.e. a crop with low biomass) should be offset with a high biomass cover crop, like a mixture of grasses and legumes.

There is no limit to the number of crops that can be used in a rotation, or the amount of time a rotation takes to complete. Decisions about rotations are made years prior, seasons prior, or even at the very last minute when an opportunity to increase profits or soil quality presents itself. In short, there is no singular formula for rotation, but many considerations to take into account.

Implementation

Crop rotation systems may be enriched by the influences of other practices such as the addition of livestock and manure, intercropping or multiple cropping, and organic management low in pesticides and synthetic fertilizers.

Incorporation of Livestock

Introducing livestock makes the most efficient use of critical sod and cover crops; livestock (through manure) are able to distribute the nutrients in these crops throughout the soil rather than removing nutrients from the farm through the sale of hay. In systems where use of farm livestock would violate reservations growers or consumers may have about animal exploitation, efforts are made to surrogate this input through livestock in the soil, namely worms and microorganisms.

In Sub-Saharan Africa, as animal husbandry becomes less of a nomadic practice many herders have begun integrating crop production into their practice. This is known as mixed farming, or the practice of crop cultivation with the incorporation of raising cattle, sheep and/or goats by the same economic entity, is increasingly common. This interaction between the animal, the land and the crops are being done on a small scale all across this region. Crop residues provide animal feed, while the animals provide manure for replenishing crop nutrients and draft power. Both processes are extremely important in this region of the world as it is expensive and logistically unfeasible to transport in synthetic fertilizers and large-scale machinery. As an additional benefit, the cattle, sheep and/or goat provide milk and can act as a cash crop in the times of economic hardship.

Organic Farming

Crop rotation is a required practice in order for a farm to receive organic certification in the United States. The "Crop Rotation Practice Standard" for the National Organic Program under the U.S. Code of Federal Regulations, section §205.205, states that:

Farmers are required to implement a crop rotation that maintains or builds soil organic matter, works to control pests, manages and conserves nutrients, and protects against erosion. Producers of perennial crops that aren't rotated may utilize other practices, such as cover crops, to maintain soil health.

In addition to lowering the need for inputs by controlling for pests and weeds and increasing available nutrients, crop rotation helps organic growers increase the amount of biodiversity on their farms. Biodiversity is also a requirement of organic certification, however, there are no rules in place to regulate or reinforce this standard. Increasing the biodiversity of crops has beneficial effects on the surrounding ecosystem and can host a greater diversity of fauna, insects, and beneficial microorganism in the soil.< Some studies point to increased nutrient availability from crop rotation under organic systems compared to conventional practices as organic practices are less likely to inhibit of beneficial microbes in soil organic matter.

While multiple cropping and intercropping benefit from many of the same principals as crop rotation, they do not satisfy the requirement under the NOP.

Intercropping

Multiple cropping systems, such as intercropping or companion planting, offer more diversity and complexity within the same season or rotation, for example the three sisters. An example of companion planting is the inter-planting of corn with pole beans and vining squash or pumpkins. In this system, the beans provide nitrogen; the corn provides support for the beans and a "screen" against squash vine borer; the vining squash provides a weed suppressive canopy and discourages corn-hungry raccoons.

Double-cropping is common where two crops, typically of different species, are grown sequentially in the same growing season, or where one crop (e.g. vegetable) is grown continuously with a cover crop (e.g. wheat). This is advantageous for small farms, who often cannot afford to leave cover crops to replenish the soil for extended periods of time, as larger farms can. When multiple cropping is implemented on small farms, these systems can maximize benefits of crop rotation on available land resources.

Benefits

Agronomists describe the benefits to yield in rotated crops as "The Rotation Effect". There are many found benefits of rotation systems: however, there is no specific scientific basis for the sometimes 10-25% yield increase in a crop grown in rotation versus monoculture. The factors related to the increase are simply described as alleviation of the negative factors of monoculture cropping systems. Explanations due to improved nutrition; pest, pathogen, and weed stress reduction; and improved soil structure have been found in some cases to be correlated, but causation has not been determined for the majority of cropping systems.

Other benefits of rotation cropping systems include production cost advantages. Overall financial risks are more widely distributed over more diverse production of crops and/or livestock. Less reliance is placed on purchased inputs and over time crops can maintain production goals with fewer inputs. This in tandem with greater short and long term yields makes rotation a powerful tool for improving agricultural systems.

Soil Organic Matter

The use of different species in rotation allows for increased soil organic matter (SOM), greater soil structure, and improvement of the chemical and biological soil environment for crops. With more SOM, water infiltration and retention improves, providing increased drought tolerance and decreased erosion.

Soil organic matter is a mix of decaying material from biomass with active microorganisms. Crop rotation, by nature, increases exposure to biomass from sod, green manure, and a various other plant debris. The reduced need for intensive tillage under crop rotation allows biomass aggregation to lead to greater nutrient retention and utilization, decreasing the need for added nutrients. With tillage, disruption and oxidation of soil creates a less conducive environment for diversity and proliferation of microorganisms in the soil. These microorganisms are what make nutrients available to plants. So, where "active" soil organic matter is a key to productive soil, soil with low microbial activity provides significantly fewer nutrients to plants; this is true even though the quantity of biomass left in the soil may be the same.

Soil microorganisms also decrease pathogen and pest activity through competition. In addition, plants produce root exudates and other chemicals which manipulate their soil environment as well as their weed environment. Thus rotation allows increased yields from nutrient availability but also alleviation of allelopathy and competitive weed environments.

Carbon Sequestration

Studies have shown that crop rotations greatly increase soil organic carbon (SOC) content, the main constituent of soil organic matter. Carbon, along with hydrogen and oxygen, is a macronutrient for plants. Highly diverse rotations spanning long periods of time have shown to be even more effective in increasing SOC, while soil disturbances (e.g. from tillage) are responsible for exponential decline in SOC levels. In Brazil, conservation to no-till methods combined with intensive crop rotations has been shown an SOC sequestration rate of 0.41 tonnes per hectare per year.

In addition to enhancing crop productivity, sequestration of atmospheric carbon has great implications in reducing rates of climate change by removing carbon dioxide from the air.

Nitrogen Fixing

Rotating crops adds nutrients to the soil. Legumes, plants of the family Fabaceae, for instance, have nodules on their roots which contain nitrogen-fixing bacteria called rhizobia. It therefore makes good sense agriculturally to alternate them with cereals (family Poaceae) and other plants that require nitrates.

Pathogen and Pest Control

Crop rotation is also used to control pests and diseases that can become established in the soil over time. The changing of crops in a sequence decreases the population level of pests by (1) interrupting pest life cycles and (2) interrupting pest habitat. Plants within the same taxonomic family tend to have similar pests and pathogens. By regularly changing crops and keeping the soil occupied by cover crops instead of lying fallow, pest cycles can be broken or limited, especially cycles that benefit from overwintering in residue. For example, root-knot nematode is a serious problem for some plants in warm climates and sandy soils, where it slowly builds up to high levels in the soil, and can severely damage plant productivity by cutting off circulation from the plant roots. Growing a crop that is not a host for root-knot nematode for one season greatly reduces the level of the nematode in the soil, thus making it possible to grow a susceptible crop the following season without needing soil fumigation.

This principle is of particular use in organic farming, where pest control must be achieved without synthetic pesticides.

Weed Management

Integrating certain crops, especially cover crops, into crop rotations is of particular value to weed management. These crops crowd out weed through competition. In addition, the sod and compost from cover crops and green manure slows the growth of what weeds are still able to make it through the soil, giving the crops further competitive advantage. By removing slowing the growth and proliferation of weeds while cover crops are cultivated, farmers greatly reduce the presence of weeds for future crops, including shallow rooted and row crops, which are less resistant to weeds. Cover crops are, therefore, considered conservation crops because they protect otherwise fallow land from becoming overrun with weeds.

This system has advantages over other common practices for weeds management, such as tillage.

Tillage is meant to inhibit growth of weeds by overturning the soil; however, this has a countering effect of exposing weed seeds that may have gotten buried and burying valuable crop seeds. Under crop rotation, the number of viable seeds in the soil is reduced through the reduction of the weed population.

Preventing Soil Erosion

Crop rotation can significantly reduce the amount of soil lost from erosion by water. In areas that are highly susceptible to erosion, farm management practices such as zero and reduced tillage can be supplemented with specific crop rotation methods to reduce raindrop impact, sediment detachment, sediment transport, surface runoff, and soil loss.

Protection against soil loss is maximized with rotation methods that leave the greatest mass of crop stubble (plant residue left after harvest) on top of the soil. Stubble cover in contact with the soil minimizes erosion from water by reducing overland flow velocity, stream power, and thus the ability of the water to detach and transport sediment. Soil Erosion and Cill prevent the disruption and detachment of soil aggregates that cause macropores to block, infiltration to decline, and runoff to increase. This significantly improves the resilience of soils when subjected to periods of erosion and stress.

The effect of crop rotation on erosion control varies by climate. In regions under relatively consistent climate conditions, where annual rainfall and temperature levels are assumed, rigid crop rotations can produce sufficient plant growth and soil cover. In regions where climate conditions are less predictable, and unexpected periods of rain and drought may occur, a more flexible approach for soil cover by crop rotation is necessary. An opportunity cropping system promotes adequate soil cover under these erratic climate conditions. In an opportunity cropping system, crops are grown when soil water is adequate and there is a reliable sowing window. This form of cropping system is likely to produce better soil cover than a rigid crop rotation because crops are only sown under optimal conditions, whereas rigid systems are not necessarily sown in the best conditions available.

Crop rotations also affect the timing and length of when a field is subject to fallow. This is very important because depending on a particular region's climate, a field could be the most vulnerable to erosion when it is under fallow. Efficient fallow management is an essential part of reducing erosion in a crop rotation system. Zero tillage is a fundamental management practice that promotes crop stubble retention under longer unplanned fallows when crops cannot be planted. Such management practices that succeed in retaining suitable soil cover in areas under fallow will ultimately reduce soil loss.

Biodiversity

Increasing the biodiversity of crops has beneficial effects on the surrounding ecosystem and can host a greater diversity of fauna, insects, and beneficial microorganisms in the soil. Some studies point to increased nutrient availability from crop rotation under organic systems compared to conventional practices as organic practices are less likely to inhibit of beneficial microbes in soil organic matter, such as arbuscular mycorrhizae, which increase nutrient uptake in plants. Increasing biodiversity also increases the resilience of agro-ecological systems.

Farm Productivity

Crop rotation contributes to increased yields through improved soil nutrition. By requiring planting and harvesting of different crops at different times, more land can be farmed with the same amount of machinery and labour.

Risk Management

Different crops in the rotation can reduce the risks of adverse weather for the individual farmer.

Challenges

While crop rotation requires a great deal of planning, crop choice must respond to a number of fixed conditions (soil type, topography, climate, and irrigation) in addition to conditions that may change dramatically from year to the next (weather, market, labor supply). In this way, it is unwise to plan crops years in advance. Improper implementation of a crop rotation plan may lead to imbalances in the soil nutrient composition or a buildup of pathogens affecting a critical crop. The consequences of faulty rotation may take years to become apparent even to experienced soil scientists and can take just as long to correct.

Many challenges exist within the practices associated with crop rotation. For example, green manure from legumes can lead to an invasion of snails or slugs and the decay from green manure can occasionally suppress the growth of other crops.

Cover Crop

A cover crop is a crop planted primarily to manage soil erosion, soil fertility, soil quality, water, weeds, pests, diseases, biodiversity and wildlife in an *agroecosystem* (Lu *et al.* 2000), an ecological system managed and largely shaped by humans across a range of intensities to produce food, feed, or fiber. Currently, not many countries are known for using the cover crop method.

Cover crops are of interest in sustainable agriculture as many of them improve the sustainability of agroecosystem attributes and may also indirectly improve qualities of neighboring natural ecosystems. Farmers choose to grow and manage specific cover crop types based on their own needs and goals, influenced by the biological, environmental, social, cultural, and economic factors of the food system in which they operate (Snapp *et al.* 2005). The farming practice of cover crops has been recognized as climate-smart agriculture by the White House.

Soil Erosion

Although cover crops can perform multiple functions in an agroecosystem simultaneously, they are often grown for the sole purpose of preventing soil erosion. Soil erosion is a process that can irreparably reduce the productive capacity of an agroecosystem. Dense cover crop stands physically slow down the velocity of rainfall before it contacts the soil surface, preventing soil splashing and erosive surface runoff (Romkens *et al.* 1990). Additionally, vast cover crop root networks help anchor the soil in place and increase soil porosity, creating suitable habitat networks for soil macrofauna (Tomlin *et al.* 1995). It keeps the enrichment of the soil good for the next few years.

Soil Fertility Management

One of the primary uses of cover crops is to increase soil fertility. These types of cover crops are referred to as "green manure." They are used to manage a range of soil macronutrients and micronutrients. Of the various nutrients, the impact that cover crops have on nitrogen management has received the most attention from researchers and farmers, because nitrogen is often the most limiting nutrient in crop production.

Often, green manure crops are grown for a specific period, and then plowed under before reaching full maturity in order to improve soil fertility and quality. Also the stalks left block the soil from being eroded.

Green manure crops are commonly leguminous, meaning they are part of the Fabaceae (pea) family.This family is unique in that all of the species in it set pods, such as bean, lentil, lupins and alfalfa. Leguminous cover crops are typically high in nitrogen and can often provide the required quantity of nitrogen for crop production. In conventional farming, this nitrogen is typically applied in chemical fertilizer form. This quality of cover crops is called fertilizer replacement value (Thiessen-Martens *et al.* 2005).

Another quality unique to leguminous cover crops is that they form symbiotic relationships with the rhizobial bacteria that reside in legume root nodules. Lupins is nodulated by the soil microorganism *Bradyrhizobium* sp. (Lupinus). Bradyrhizobia are encountered as microsymbionts in other leguminous crops (*Argyrolobium, Lotus, Ornithopus, Acacia, Lupinus*) of Mediterranean origin. These bacteria convert biologically unavailable atmospheric nitrogen gas (N) to biologically available ammonium (NH+4) through the process of biological nitrogen fixation.

Prior to the advent of the Haber-Bosch process, an energy-intensive method developed to carry out industrial nitrogen fixation and create chemical nitrogen fertilizer, most nitrogen introduced to ecosystems arose through biological nitrogen fixation (Galloway *et al.* 1995). Some scientists believe that widespread biological nitrogen fixation, achieved mainly through the use of cover crops, is the only alternative to industrial nitrogen fixation in the effort to maintain or increase future food production levels (Bohlool *et al.* 1992, Peoples and Craswell 1992, Giller and Cadisch 1995). Industrial nitrogen fixation has been criticized as an unsustainable source of nitrogen for food production due to its reliance on fossil fuel energy and the environmental impacts associated with chemical nitrogen fertilizer use in agriculture (Jensen and Hauggaard-Nielsen 2003). Such widespread environmental impacts include nitrogen fertilizer losses into waterways, which can lead to eutrophication (nutrient loading) and ensuing hypoxia (oxygen depletion) of large bodies of water.

An example of this lies in the Mississippi Valley Basin, where years of fertilizer nitrogen loading into the watershed from agricultural production have resulted in a hypoxic "dead zone" off the Gulf of Mexico the size of New Jersey (Rabalais *et al.* 2002). The ecological complexity of marine life in this zone has been diminishing as a consequence (CENR 2000).

As well as bringing nitrogen into agroecosystems through biological nitrogen fixation, types of cover crops known as "catch crops" are used to retain and recycle soil nitrogen already present. The catch crops take up surplus nitrogen remaining from fertilization of the previous crop, preventing it from being lost through leaching (Morgan *et al.* 1942), or gaseous denitrification or volatilization (Thorup-Kristensen *et al.* 2003).

Catch crops are typically fast-growing annual cereal species adapted to scavenge available nitrogen efficiently from the soil (Ditsch and Alley 1991). The nitrogen tied up in catch crop biomass is released back into the soil once the catch crop is incorporated as a green manure or otherwise begins to decompose.

An example of green manure use comes from Nigeria, where the cover crop *Mucuna pruriens* (velvet bean) has been found to increase the availability of phosphorus in soil after a farmer applies rock phosphate (Vanlauwe *et al.* 2000).

Soil Quality Management

Cover crops can also improve soil quality by increasing soil organic matter levels through the input of cover crop biomass over time. Increased soil organic matter enhances soil structure, as well as the water and nutrient holding and buffering capacity of soil (Patrick *et al.* 1957). It can also lead to increased soil carbon sequestration, which has been promoted as a strategy to help offset the rise in atmospheric carbon dioxide levels (Kuo *et al.* 1997, Sainju *et al.* 2002, Lal 2003).

Soil quality is managed to produce optimum circumstances for crops to flourish. The principal factors of soil quality are soil salination, pH, microorganism balance and the prevention of soil contamination.

Water Management

By reducing soil erosion, cover crops often also reduce both the rate and quantity of water that drains off the field, which would normally pose environmental risks to waterways and ecosystems downstream (Dabney *et al.* 2001). Cover crop biomass acts as a physical barrier between rainfall and the soil surface, allowing raindrops to steadily trickle down through the soil profile. Also, as stated above, cover crop root growth results in the formation of soil pores, which in addition to enhancing soil macrofauna habitat provides pathways for water to filter through the soil profile rather than draining off the field as surface flow. With increased water infiltration, the potential for soil water storage and the recharging of aquifers can be improved (Joyce *et al.* 2002).

Just before cover crops are killed (by such practices including mowing, tilling, discing, rolling, or herbicide application) they contain a large amount of moisture. When the cover crop is incorporated into the soil, or left on the soil surface, it often increases soil moisture. In agroecosystems where water for crop production is in short supply, cover crops can be used as a mulch to conserve water by shading and cooling the soil surface. This reduces evaporation of soil moisture. In other situations farmers try to dry the soil out as quickly as possible going into the planting season. Here prolonged soil moisture conservation can be problematic.

While cover crops can help to conserve water, in temperate regions (particularly in years with below average precipitation) they can draw down soil water supply in the spring, particularly if climatic growing conditions are good. In these cases, just before crop planting, farmers often face a tradeoff between the benefits of increased cover crop growth and the drawbacks of reduced soil moisture for cash crop production that season. C/N ratio is balanced with this application.

Weed Management

Cover crop in South Dakota

Thick cover crop stands often compete well with weeds during the cover crop growth period, and can prevent most germinated weed seeds from completing their life cycle and reproducing. If the cover crop is left on the soil surface rather than incorporated into the soil as a green manure after its growth is terminated, it can form a nearly impenetrable mat. This drastically reduces light transmittance to weed seeds, which in many cases reduces weed seed germination rates (Teasdale 1993). Furthermore, even when weed seeds germinate, they often run out of stored energy for growth before building the necessary structural capacity to break through the cover crop mulch layer. This is often termed the *cover crop smother effect* (Kobayashi *et al.* 2003).

Some cover crops suppress weeds both during growth and after death (Blackshaw *et al.* 2001). During growth these cover crops compete vigorously with weeds for available space, light, and nutrients, and after death they smother the next flush of weeds by forming a mulch layer on the soil surface. For example, Blackshaw *et al.* (2001) found that when using *Melilotus officinalis* (yellow sweetclover) as a cover crop in an improved fallow system (where a fallow period is intentionally improved by any number of different management practices, including the planting of cover crops), weed biomass only constituted between 1-12% of total standing biomass at the end of the cover crop growing season. Furthermore, after cover crop termination, the yellow sweetclover residues suppressed weeds to levels 75-97% lower than in fallow (no yellow sweetclover) systems .

In addition to competition-based or physical weed suppression, certain cover crops are known to suppress weeds through allelopathy (Creamer *et al.* 1996, Singh *et al.* 2003). This occurs when certain biochemical cover crop compounds are degraded that happen to be toxic to, or inhibit seed germination of, other plant species. Some well known examples of allelopathic cover crops are *Secale cereale* (rye), *Vicia villosa* (hairy vetch), *Trifolium pratense* (red clover), *Sorghum bicolor* (sorghum-sudangrass), and species in the Brassicaceae family, particularly mustards (Haramoto and Gallandt 2004). In one study, rye cover crop residues were found to have provided between 80% and 95% control of early season broadleaf weeds when used as a mulch during the production of different cash crops such as soybean, tobacco, corn, and sunflower (Nagabhushana *et al.* 2001).

In a recent study released by the Agricultural Research Service (ARS) scientists examined how rye seeding rates and planting patterns affected cover crop production. The results show that planting more pounds per acre of rye increased the cover crop's production as well as decreased the amount of weeds. The same was true when scientists tested seeding rates on legumes and oats; a higher density of seeds planted per acre decreased the amount of weeds and increased the yield of legume and oat production. The planting patterns, which consisted of either traditional rows or grid patterns, did not seem to make a significant impact on the cover crop's production or on the weed production in either cover crop. The ARS scientists concluded that increased seeding rates could be an effective method of weed control.

Disease Management

In the same way that allelopathic properties of cover crops can suppress weeds, they can also break disease cycles and reduce populations of bacterial and fungal diseases (Everts 2002), and parasitic nematodes (Potter *et al.* 1998, Vargas-Ayala *et al.* 2000). Species in the Brassicaceae family, such as mustards, have been widely shown to suppress fungal disease populations through the release of naturally occurring toxic chemicals during the degradation of glucosinolade compounds in their plant cell tissues (Lazzeri and Manici 2001).

Pest Management

Some cover crops are used as so-called "trap crops", to attract pests away from the crop of value and toward what the pest sees as a more favorable habitat (Shelton and Badenes-Perez 2006). Trap crop areas can be established within crops, within farms, or within landscapes. In many cases the trap crop is grown during the same season as the food crop being produced. The limited area occupied by these trap crops can be treated with a pesticide once pests are drawn to the trap in large enough numbers to reduce the pest populations. In some organic systems, farmers drive over the trap crop with a large vacuum-based implement to physically pull the pests off the plants and out of the field (Kuepper and Thomas 2002). This system has been recommended for use to help control the lygus bugs in organic strawberry production (Zalom *et al.* 2001). Another example of trap crops are nematode resistance White mustard (*Sinapis alba*) and Radish (*Raphanus sativus*). They can be grown after a main (cereal) crop and trap nematodes, for example the beet cyst nematode and Columbian root knot nematode. When grown, nematodes hatch and are attracted to the roots. After entering the roots they cannot reproduce in the root due to a hypersensitive resistance reaction of the plant. Hence the nematode population is greatly reduced, by 70-99%, depending on species and cultivation time.

Other cover crops are used to attract natural predators of pests by providing elements of their habitat. This is a form of biological control known as habitat augmentation, but achieved with the use of cover crops (Bugg and Waddington 1994). Findings on the relationship between cover crop presence and predator/pest population dynamics have been mixed, pointing toward the need for detailed information on specific cover crop types and management practices to best complement a given integrated pest management strategy. For example, the predator mite *Euseius tularensis* (Congdon) is known to help control the pest citrus thrips in Central California citrus orchards. Researchers found that the planting of several different leguminous cover crops (such as bell bean, woollypod vetch, New Zealand white clover, and Austrian winter pea) provided sufficient pollen as a feeding source to cause a seasonal increase in *E. tularensis* populations, which with good timing could potentially introduce enough predatory pressure to reduce pest populations of citrus thrips (Grafton-Cardwell *et al.* 1999).

Diversity and wildlife

Although cover crops are normally used to serve one of the above discussed purposes, they often simultaneously improve farm habitat for wildlife. The use of cover crops adds at least one more dimension of plant diversity to a cash crop rotation. Since the cover crop is typically not a crop of value, its management is usually less intensive, providing a window of "soft" human influence on the farm. This relatively "hands-off" management, combined with the increased on-farm heterogeneity created by the establishment of cover crops, increases the likelihood that a more complex trophic structure will develop to support a higher level of wildlife diversity (Freemark and Kirk 2001).

In one study, researchers compared arthropod and songbird species composition and field use between conventionally and cover cropped cotton fields in the Southern United States. The cover cropped cotton fields were planted to clover, which was left to grow in between cotton rows throughout the early cotton growing season (stripcover cropping). During the migration and breeding season, they found that songbird densities were 7–20 times higher in the cotton fields with integrated clover cover crop than in the conventional cotton fields. Arthropod abundance and biomass was also higher in the clover cover cropped fields throughout much of the songbird breeding season, which was attributed to an increased supply of flower nectar from the clover. The clover cover crop enhanced songbird habitat by providing cover and nesting sites, and an increased food source from higher arthropod populations (Cederbaum *et al.* 2004).

Tillage

Cultivating after an early rain

Tillage is the agricultural preparation of soil by mechanical agitation of various types, such as digging, stirring, and overturning. Examples of human-powered tilling methods using hand tools include shovelling, picking, mattock work, hoeing, and raking. Examples of draft-animal-powered or mechanized work include ploughing (overturning with moldboards or chiseling with chisel shanks), rototilling, rolling with cultipackers or other rollers, harrowing, and cultivating with cultivator shanks (teeth). Small-scale gardening and farming, for household food production or small business production, tends to use the smaller-scale methods above, whereas medium- to large-scale farming tends to use the larger-scale methods. There is a fluid continuum, however.

Any type of gardening or farming, but especially larger-scale commercial types, may also use low-till or no-till methods as well.

Tillage is often classified into two types, primary and secondary. There is no strict boundary between them so much as a loose distinction between tillage that is deeper and more thorough (primary) and tillage that is shallower and sometimes more selective of location (secondary). Primary tillage such as ploughing tends to produce a rough surface finish, whereas secondary tillage tends to produce a smoother surface finish, such as that required to make a good seed-bed for many crops. Harrowing and rototilling often combine primary and secondary tillage into one operation.

"Tillage" can also mean the land that is tilled. The word "cultivation" has several senses that overlap substantially with those of "tillage". In a general context, both can refer to agriculture. Within agriculture, both can refer to any of the kinds of soil agitation described above. Additionally, "cultivation" or "cultivating" may refer to an even narrower sense of shallow, selective secondary tillage of row crop fields that kills weeds while sparing the crop plants.

Tillage Systems

Reduced Tillage

Plough tilling the field

Reduced tillage leaves between 15 and 30% residue cover on the soil or 500 to 1000 pounds per acre (560 to 1100 kg/ha) of small grain residue during the critical erosion period. This may involve the use of a chisel plow, field cultivators, or other implements.

Intensive Tillage

Intensive tillage leaves less than 15% crop residue cover or less than 500 pounds per acre (560 kg/ha) of small grain residue. This type of tillage is often referred to as conventional tillage but as conservational tillage is now more widely used than intensive tillage (in the United States), it is often not appropriate to refer to this type of tillage as conventional. Intensive tillage often involves multiple operations with implements such as a mold board, disk, and/or chisel plow. Then a finisher with a harrow, rolling basket, and cutter can be used to prepare the seed bed. There are many variations.

Conservation Tillage

Conservation tillage leaves at least 30% of crop residue on the soil surface, or at least 1,000 lb/ac (1,100 kg/ha) of small grain residue on the surface during the critical soil erosion period. This slows water movement, which reduces the amount of soil erosion. Conservation tillage also benefits farmers by reducing fuel consumption and soil compaction. By reducing the number of times the farmer travels over the field, farmers realize significant savings in fuel and labor. In most years since 1997, conservation tillage was used in US cropland more than intensive or reduced tillage.

However, conservation tillage delays warming of the soil due to the reduction of dark earth exposure to the warmth of the spring sun, thus delaying the planting of the next year's spring crop of corn.

- No-till - Never use a plow, disk, etc. ever again. Aims for 100% ground cover.

- Strip-Till - Narrow strips are tilled where seeds will be planted, leaving the soil in between the rows untilled.

- Mulch-till

- Rotational Tillage - Tilling the soil every two years or less often (every other year, or every third year, etc.).

- Ridge-Till

- Zone tillage - This form of conservation tillage is further explained below.

Zone Tillage

Zone tillage is a form of modified deep tillage in which only narrow strips are tilled, leaving soil in between the rows untilled. This type of tillage agitates the soil to help reduce soil compaction problems and to improve internal soil drainage.

Purpose

Zone tillage is designed to only disrupt the soil in a narrow strip directly below the crop row. In comparison to no-till, which relies on the previous year's plant residue to protect the soil and aides in postponement of the warming of the soil and crop growth in Northern climates, zone tillage creates approximately a 5-inch-wide strip that simultaneously breaks up plow pans, assists in warming the soil and helps to prepare a seedbed. When combined with cover crops, zone tillage helps replace lost organic matter, slows the deterioration of the soil, improves soil drainage, increases soil water and nutrient holding capacity, and allows necessary soil organisms to survive.

Usage

It has been successfully used on farms in the mid-west and west for over 40 years and is currently used on more than 36% of the U.S. farmland. Some specific states where zone tillage is currently in practice are Pennsylvania, Connecticut, Minnesota, Indiana, Wisconsin, and Illinois.

Unfortunately, there aren't consistent yield results in the Northern Cornbelt states; however, there is still interest in deep tillage within the agriculture industry. In areas that are not well-drained, deep tillage may be used as an alternative to installing more expensive tile drainage.

Effects of Tillage

Positive

Plowing:

- Loosens and aerates the top layer of soil, which facilitates planting the crop

- Helps mix harvest residue, organic matter (humus), and nutrients evenly into the soil

- Mechanically destroys weeds

- Dries the soil before seeding (in wetter climates tillage aids in keeping the soil drier)

- When done in autumn, helps exposed soil crumble over winter through frosting and defrosting, which helps prepare a smooth surface for spring planting

Negative

- Dries the soil before seeding

- Soil loses a lot of nutrients, like nitrogen and fertilizer, and its ability to store water

- Decreases the water infiltration rate of soil. (Results in more runoff and erosion since the soil absorbs water slower than before)

- Tilling the soil results in dislodging the cohesiveness of the soil particles thereby inducing erosion.

- Chemical runoff

- Reduces organic matter in the soil

- Reduces microbes, earthworms, ants, etc.

- Destroys soil aggregates

- Compaction of the soil, also known as a tillage pan

- Eutrophication (nutrient runoff into a body of water)

- Can attract slugs, cut worms, army worms, and harmful insects to the left over residues.

- Crop diseases can be harbored in surface residues

General Comments

- The type of implement makes the most difference, although other factors can have an effect.

- Tilling in absolute darkness (night tillage) might reduce the number of weeds that sprout following the tilling operation by half. Light is necessary to break the dormancy of some weed species' seed, so if fewer seeds are exposed to light during the tilling process, fewer will sprout. This may help reduce the amount of herbicides needed for weed control.

- Greater speeds, when using certain tillage implements (disks and chisel plows), lead to more intensive tillage (i.e., less residue is on the soil surface).

- Increasing the angle of disks causes residues to be buried more deeply. Increasing their concavity makes them more aggressive.

- Chisel plows can have spikes or sweeps. Spikes are more aggressive.

- Percentage residue is used to compare tillage systems because the amount of crop residue affects the soil loss due to erosion.

- Soybean management practices to see what types of tillage are currently recommended for Soybean Production.

Definitions

Primary tillage loosens the soil and mixes in fertilizer and/or plant material, resulting in soil with a rough texture.

Secondary tillage produces finer soil and sometimes shapes the rows, preparing the seed bed. It also provides weed control throughout the growing season during the maturation of the crop plants, unless such weed control is instead achieved with low-till or no-till methods involving herbicides.

- The seed bed preparation can be done with harrows (of which there are many types and subtypes), dibbles, hoes, shovels, rotary tillers, subsoilers, ridge- or bed-forming tillers, rollers, or cultivators.

- The weed control, to the extent that it is done via tillage, is usually achieved with cultivators or hoes, which disturb the top few centimeters of soil around the crop plants but with minimal disturbance of the crop plants themselves. The tillage kills the weeds via 2 mechanisms: uprooting them, burying their leaves (cutting off their photosynthesis), or a combination of both. Weed control both prevents the crop plants from being outcompeted by the weeds (for water and sunlight) and prevents the weeds from reaching their seed stage, thus reducing future weed population aggressiveness.

History of Tilling

Tilling was first performed via human labor, sometimes involving slaves. Hoofed animals could also be used to till soil via trampling. The wooden plow was then invented. It could be pulled by mule, ox, elephant, water buffalo, or similar sturdy animal. Horses are generally unsuitable, though breeds such as the scyne could work. The steel plow allowed farming in the American Midwest, where tough prairie grasses and rocks caused trouble. Soon after 1900, the farm tractor was introduced, which eventually made modern large-scale agriculture possible.

Tilling with Hungarian Grey cattles

Alternatives to Tilling

Modern agricultural science has greatly reduced the use of tillage. Crops can be grown for several years without any tillage through the use of herbicides to control weeds, crop varieties that tolerate packed soil, and equipment that can plant seeds or fumigate the soil without really digging it up. This practice, called no-till farming, reduces costs and environmental change by reducing soil erosion and diesel fuel usage.

Site Preparation of Forest Land

Site preparation is any of various treatments applied to a site in order to ready it for seeding or planting. The purpose is to facilitate the regeneration of that site by the chosen method. Site preparation may be designed to achieve, singly or in any combination: improved access, by reducing or rearranging slash, and amelioration of adverse forest floor, soil, vegetation, or other biotic factors. Site preparation is undertaken to ameliorate one or more constraints that would otherwise be likely to thwart the objectives of management. A valuable bibliography on the effects of soil temperature and site preparation on subalpine and boreal tree species has been prepared by McKinnon et al. (2002).

Site preparation is the work that is done before a forest area is regenerated. Some types of site preparation are burning.

Burning

Broadcast burning is commonly used to prepare clearcut sites for planting, e.g., in central British Columbia, and in the temperate region of North America generally.

Prescribed burning is carried out primarily for slash hazard reduction and to improve site conditions for regeneration; all or some of the following benefits may accrue:

a) Reduction of logging slash, plant competition, and humus prior to direct seeding, planting, scarifying or in anticipation of natural seeding in partially cut stands or in connection with seed-tree systems.

b) Reduction or elimination of unwanted forest cover prior to planting or seeding, or prior to preliminary scarification thereto.

c) Reduction of humus on cold, moist sites to favour regeneration.

d) Reduction or elimination of slash, grass, or brush fuels from strategic areas around forested land to reduce the chances of damage by wildfire.

Prescribed burning for preparing sites for direct seeding was tried on a few occasions in Ontario, but none of the burns was hot enough to produce a seedbed that was adequate without supplementary mechanical site preparation.

Changes in soil chemical properties associated with burning include significantly increased pH, which Macadam (1987) in the Sub-boreal Spruce Zone of central British Columbia found persisting more than a year after the burn. Average fuel consumption was 20 to 24 t/ha and the forest floor depth was reduced by 28% to 36%. The increases correlated well with the amounts of slash (both total and ≥7 cm diameter) consumed. The change in pH depends on the severity of the burn and the amount consumed; the increase can be as much as 2 units, a 100-fold change. Deficiencies of copper and iron in the foliage of white spruce on burned clearcuts in central British Columbia might be attributable to elevated pH levels.

Even a broadcast slash fire in a clearcut does not give a uniform burn over the whole area. Tarrant (1954), for instance, found only 4% of a 140-ha slash burn had burned severely, 47% had burned lightly, and 49% was unburned. Burning after windrowing obviously accentuates the subsequent heterogeneity.

Marked increases in exchangeable calcium also correlated with the amount of slash at least 7 cm in diameter consumed. Phosphorus availability also increased, both in the forest floor and in the 0 cm to 15 cm mineral soil layer, and the increase was still evident, albeit somewhat diminished, 21 months after burning. However, in another study in the same Sub-boreal Spruce Zone found that although it increased immediately after the burn, phosphorus availability had dropped to below pre-burn levels within 9 months.

Nitrogen will be lost from the site by burning, though concentrations in remaining forest floor were found by Macadam (1987) to have increased in 2 of 6 plots, the others showing decreases. Nutrient losses may be outweighed, at least in the short term, by improved soil microclimate through the reduced thickness of forest floor where low soil temperatures are a limiting factor.

The *Picea/Abies* forests of the Alberta foothills are often characterized by deep accumulations of organic matter on the soil surface and cold soil temperatures, both of which make reforestation difficult and result in a general deterioration in site productivity; Endean and Johnstone (1974) describe experiments to test prescribed burning as a means of seedbed preparation and site amelioration on representative clear-felled *Picea/Abies* areas. Results showed that, in general, prescribed burning did not reduce organic layers satisfactorily, nor did it increase soil temperature, on the sites tested. Increases in seedling establishment, survival, and growth on the burned sites were probably the result of slight reductions in the depth of the organic layer, minor increases in soil temperature, and marked improvements in the efficiency of the planting crews. Results also suggested that the process of site deterioration has not been reversed by the burning treatments applied.

Ameliorative Intervention

Slash weight (the oven-dry weight of the entire crown and that portion of the stem < 4 inches in diameter) and size distribution are major factors influencing the forest fire hazard on harvested sites. Forest managers interested in the application of prescribed burning for hazard reduction and silviculture, were shown a method for quantifying the slash load by Kiil (1968). In west-central Alberta, he felled, measured, and weighed 60 white spruce, graphed (a) slash weight per merchantable unit volume against diameter at breast height (dbh), and (b) weight of fine slash (<1.27 cm) also against dbh, and produced a table of slash weight and size distribution on one acre of a hypothetical stand of white spruce. When the diameter distribution of a stand is unknown, an estimate of slash weight and size distribution can be obtained from average stand diameter, number of trees per unit area, and merchantable cubic foot volume. The sample trees in Kiil's study had full symmetrical crowns. Densely growing trees with short and often irregular crowns would probably be overestimated; open-grown trees with long crowns would probably be underestimated.

The need to provide shade for young outplants of Engelmann spruce in the high Rocky Mountains is emphasized by the U.S. Forest Service. Acceptable planting spots are defined as microsites on the north and east sides of down logs, stumps, or slash, and lying in the shadow cast by such material. Where the objectives of management specify more uniform spacing, or higher densities, than obtainable from an existing distribution of shade-providing material, redistribution or importing of such material has been undertaken.

Access

Site preparation on some sites might be done simply to facilitate access by planters, or to improve access and increase the number or distribution of microsites suitable for planting or seeding.

Wang et al. (2000) determined field performance of white and black spruces 8 and 9 years after outplanting on boreal mixedwood sites following site preparation (Donaren disc trenching versus no trenching) in 2 plantation types (open versus sheltered) in southeastern Manitoba. Donaren trenching slightly reduced the mortality of black spruce but significantly increased the mortality of white spruce. Significant difference in height was found between open and sheltered plantations for black spruce but not for white spruce, and root collar diameter in sheltered plantations was significantly larger than in open plantations for black spruce but not for white spruce. Black spruce open plantation had significantly smaller volume (97 cm³) compared with black spruce sheltered (210 cm³), as well as white spruce open (175 cm³) and sheltered (229 cm³) plantations. White spruce open plantations also had smaller volume than white spruce sheltered plantations. For transplant stock, strip plantations had a significantly higher volume (329 cm³) than open plantations (204 cm³). Wang et al. (2000) recommended that sheltered plantation site preparation should be used.

Mechanical

Up to 1970, no "sophisticated" site preparation equipment had become operational in Ontario, but the need for more efficacious and versatile equipment was increasingly recognized. By this time, improvements were being made to equipment originally developed by field staff, and field testing of equipment from other sources was increasing.

According to J. Hall (1970), in Ontario at least, the most widely used site preparation technique was post-harvest mechanical scarification by equipment front-mounted on a bulldozer (blade, rake, V-plow, or teeth), or dragged behind a tractor (Imsett or S.F.I. scarifier, or rolling chopper). Drag type units designed and constructed by Ontario's Department of Lands and Forests used anchor chain or tractor pads separately or in combination, or were finned steel drums or barrels of various sizes and used in sets alone or combined with tractor pad or anchor chain units.

J. Hall's (1970) report on the state of site preparation in Ontario noted that blades and rakes were found to be well suited to post-cut scarification in tolerant hardwood stands for natural regeneration of yellow birch. Plows were most effective for treating dense brush prior to planting, often in conjunction with a planting machine. Scarifying teeth, e.g., Young's teeth, were sometimes used to prepare sites for planting, but their most effective use was found to be preparing sites for seeding, particularly in backlog areas carrying light brush and dense herbaceous growth. Rolling choppers found application in treating heavy brush but could be used only on stone-free soils. Finned drums were commonly used on jack pine–spruce cutovers on fresh brushy sites with a deep duff layer and heavy slash, and they needed to be teamed with a tractor pad unit to secure good distribution of the slash. The S.F.I. scarifier, after strengthening, had been "quite successful" for 2 years, promising trials were under way with the cone scarifier and barrel ring scarifier, and development had begun on a new flail scarifier for use on sites with shallow, rocky soils. Recognition of the need to become more effective and efficient in site preparation led the Ontario Department of Lands and Forests to adopt the policy of seeking and obtaining for field testing new equipment from Scandinavia and elsewhere that seemed to hold promise for Ontario conditions, primarily in the north. Thus, testing was begun of the Brackekultivator from Sweden and the Vako-Visko rotary furrower from Finland.

Mounding

Site preparation treatments that create raised planting spots have commonly improved outplant performance on sites subject to low soil temperature and excess soil moisture. Mounding can certainly have a big influence on soil temperature. Draper et al. (1985), for instance, documented this as well as the effect it had on root growth of outplants (Table 30).

The mounds warmed up quickest, and at soil depths of 0.5 cm and 10 cm averaged 10 and 7 °C higher, respectively, than in the control. On sunny days, daytime surface temperature maxima on the mound and organic mat reached 25 °C to 60 °C, depending on soil wetness and shading. Mounds reached mean soil temperatures of 10 °C at 10 cm depth 5 days after planting, but the control did not reach that temperature until 58 days after planting. During the first growing season, mounds had 3 times as many days with a mean soil temperature greater than 10 °C than did the control microsites.

Draper et al.'s (1985) mounds received 5 times the amount of photosynthetically active radiation (PAR) summed over all sampled microsites throughout the first growing season; the control treatment consistently received about 14% of daily background PAR, while mounds received over 70%. By November, fall frosts had reduced shading, eliminating the differential. Quite apart from its effect on temperature, incident radiation is also important photosynthetically. The average control microsite was exposed to levels of light above the compensation point for only 3 hours, i.e., one-quarter of the daily light period, whereas mounds received light above the compensation point for 11 hours, i.e., 86% of the same daily period. Assuming that incident light in the 100-600 µEm^{-2}s^{-1} intensity range is the most important for photosynthesis, the mounds received over 4 times the total daily light energy that reached the control microsites.

Orientation of Linear Site Preparation, e.g., Disk-Trenching

With linear site preparation, orientation is sometimes dictated by topography or other considerations, but the orientation can often be chosen. It can make a difference. A disk-trenching experiment in the Sub-boreal Spruce Zone in interior British Columbia investigated the effect on growth of young outplants (lodgepole pine) in 13 microsite planting positions: berm, hinge, and trench in each of north, south, east, and west aspects, as well as in untreated locations between the furrows. Tenth-year stem volumes of trees on south, east, and west-facing microsites were significantly greater than those of trees on north-facing and untreated microsites. However, planting spot selection was seen to be more important overall than trench orientation.

In a Minnesota study, the N–S strips accumulated more snow but snow melted faster than on E–W strips in the first year after felling. Snow-melt was faster on strips near the centre of the strip-felled area than on border strips adjoining the intact stand. The strips, 50 feet (15.24 m) wide, alternating with uncut strips 16 feet (4.88 m) wide, were felled in a *Pinus resinosa* stand, aged 90 to 100 years.

Windbreak

Aerial view of field windbreaks in North Dakota

One of the original buildings at Svappavaara, designed by Ralph Erskine, which forms a long windbreak.

A windbreak or shelterbelt is a plantation usually made up of one or more rows of trees or shrubs planted in such a manner as to provide shelter from the wind and to protect soil from erosion.

They are commonly planted around the edges of fields on farms. If designed properly, windbreaks around a home can reduce the cost of heating and cooling and save energy. Windbreaks are also planted to help keep snow from drifting onto roadways and even yards. Other benefits include providing habitat for wildlife and in some regions the trees are harvested for wood products.

Windbreaks and intercropping can be combined in a farming practice referred to as alleycropping. Fields are planted in rows of different crops surrounded by rows of trees. These trees provide fruit, wood, or protect the crops from the wind. Alley cropping has been particularly successful in India, Africa, and Brazil, where coffee growers have combined farming and forestry.

A further use for a shelterbelt is to screen a farm from a main road or motorway. This improves the farm landscape by reducing the visual incursion of the motorway, mitigating noise from the traffic and providing a safe barrier between farm animals and the road.

The term "windbreak" is also used to describe an article of clothing worn to prevent wind chill. Americans tend to use the term "windbreaker" whereas Europeans favor the term "windbreak".

Fences called "windbreaks" are also used. Normally made from cotton, nylon, canvas, and recycled sails, windbreaks tend to have three or more panels held in place with poles that slide into pockets sewn into the panel. The poles are then hammered into the ground and a windbreak is formed. Windbreaks or "wind fences" are used to reduce wind speeds over erodible areas such as open fields, industrial stockpiles, and dusty industrial operations. As erosion is proportional to wind speed cubed a reduction of wind speed of 1/2 (for example) will reduce erosion by over 80%.

Windbreak Aerodynamics

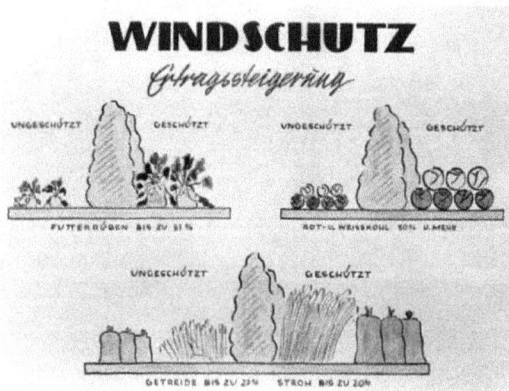

An East German windbreak promotion poster, 1952

In essence, when the wind encounters a porous obstacle such as a windbreak or shelterbelt, air pressure increases (loosely speaking, air *piles up*) on the windward side and (conversely) air pressure decreases on the leeward side. As a result, the airstream approaching the barrier is retarded, and a proportion of it is displaced up and over the barrier, resulting in a *jet* of higher wind speed aloft. The remainder of the impinging airstream, having been retarded in its approach, now circulates through the barrier to its downstream edge, pushed along by the decrease in pressure across the shelterbelt's width; emerging on the downwind side, that airstream is now further retarded by an adverse pressure gradient, because in the lee of the barrier, with increasing downwind distance air pressure *recovers* again to the ambient level. The result is that minimum wind speed occurs not at or within the windbreak, nor at its downwind edge, but further downwind - nominally, at a distance of about 3 to 5 times the windbreak height H. Beyond that point wind speed recovers, aided by downward momentum transport from the overlying, faster-moving stream. From the perspective of the Reynolds-averaged Navier–Stokes equations these effects can be understood as resulting from the loss of momentum caused by the drag of leaves and branches and would be represented by the body force f_i (a distributed momentum sink).

Not only is the mean (average) wind speed reduced in the lee of the shelter, the wind is also less gusty, for turbulent wind fluctuations are also damped. As a result, turbulent vertical mixing is weaker in the lee of the barrier than it is upwind, and interesting secondary microclimatic effects result. For instance, by day sensible heat rising from the ground due to the absorption of sunlight is mixed upward less efficiently in the lee of a windbreak, with the result that air temperature near ground is somewhat higher in the lee than on the windward side. Of course this effect is attenuated with increasing downwind distance and indeed, beyond about $8H$ downstream a zone may exist that is actually *cooler* than upwind.

Fertilizer

A large, modern fertilizer spreader

A Lite-Trac Agri-Spread lime and fertilizer spreader at an agricultural show

A fertilizer (American English) or fertiliser (British English) is any material of natural or synthetic origin (other than liming materials) that is applied to soils or to plant tissues (usually leaves) to supply one or more plant nutrients essential to the growth of plants.

Mechanism

Six tomato plants grown with and without nitrate fertilizer on nutrient-poor sand/clay soil. One of the plants in the nutrient-poor soil has died.

Fertilizers enhance the growth of plants. This goal is met in two ways, the traditional one being additives that provide nutrients. The second mode by some fertilizers act is to enhance the effectiveness of the soil by modifying its water retention and aeration. Fertilizers typically provide, in varying proportions:

- three main macronutrients:

    o   Nitrogen (N): leaf growth;

    o   Phosphorus (P): Development of roots, flowers, seeds, fruit;

    o   Potassium (K): Strong stem growth, movement of water in plants, promotion of flowering and fruiting;

- three secondary macronutrients: calcium (Ca), magnesium (Mg), and sulphur (S);

- micronutrients: copper (Cu), iron (Fe), manganese (Mn), molybdenum (Mo), zinc (Zn), boron (B), and of occasional significance there are silicon (Si), cobalt (Co), and vanadium (V) plus rare mineral catalysts.

The nutrients required for healthy plant life are classified according to the elements, but the elements are not used as fertilizers. Instead compounds containing these elements are the basis of fertilisers. The macronutrients are consumed in larger quantities and are present in plant tissue in quantities from 0.15% to 6.0% on a dry matter (DM) (0% moisture) basis. Plants are made up of four main elements: hydrogen, oxygen, carbon, and nitrogen. Carbon, hydrogen and oxygen are widely available as water and carbon dioxide. Although nitrogen makes up most of the atmosphere, it is in a form that is unavailable to plants. Nitrogen is the most important fertilizer since nitrogen is present in proteins, DNA and other components (e.g., chlorophyll). To be nutritious to plants, nitrogen must be made available in a "fixed" form. Only some bacteria and their host plants (notably legumes) can fix atmospheric nitrogen (N_2) by converting it to ammonia. Phosphate is required for the production of DNA and ATP, the main energy carrier in cells, as well as certain lipids.

Micronutrients are consumed in smaller quantities and are present in plant tissue on the order of parts-per-million (ppm), ranging from 0.15 to 400 ppm DM, or less than 0.04% DM. These elements are often present at the active sites of enzymes that carry out the plant's metabolism. Because these elements enable catalysts (enzymes) their impact far exceeds their weight percentage.

Classification

Fertilizers are classified in several ways. They are classified according to whether they provide a single nutrient (say, N, P, or K), in which case they are classified as "straight fertilizers." "Multinutrient fertilizers" (or "complex fertilizers") provide two or more nutrients, for example N and P. Fertilizers are also sometimes classified as inorganic versus organic. Inorganic fertilizers exclude carbon-containing materials except ureas. Organic fertilizers are usually (recycled) plant- or animal-derived matter. Inorganic are sometimes called synthetic fertilizers since various chemical treatments are required for their manufacture.

Single Nutrient ("Straight") Fertilizers

The main nitrogen-based straight fertilizer is ammonia or its solutions. Ammonium nitrate (NH_4NO_3) is also widely used. About 15M tons were produced in 1981. Urea is another popular source of nitrogen, having the advantage that it is a solid and non-explosive, unlike ammonia and ammonium nitrate, respectively. A few percent of the nitrogen fertilizer market (4% in 2007) has been met by calcium ammonium nitrate ($Ca(NO_3)_2 \cdot NH_4NO_3 \cdot 10H_2O$).

The main straight phosphate fertilizers are the superphosphates. "Single superphosphate" (SSP) consists of 14–18% P_2O_5, again in the form of $Ca(H_2PO_4)_2$, but also phosphogypsum ($CaSO_4 \cdot 2 H_2O$). Triple superphosphate (TSP) typically consists of 44-48% of P_2O_5 and no gypsum. A mixture of single superphosphate and triple superphosphate is called double superphosphate. More than 90% of a typical superphosphate fertilizer is water-soluble.

Multinutrient Fertilizers

These fertilizers are the most common. They consist of two or more nutrient components.

Binary (NP, NK, PK) Fertilizers

Major two-component fertilizers provide both nitrogen and phosphorus to the plants. These are called NP fertilizers. The main NP fertilizers are monoammonium phosphate (MAP) and diammonium phosphate (DAP). The active ingredient in MAP is $NH_4H_2PO_4$. The active ingredient in DAP is $(NH_4)_2HPO_4$. About 85% of MAP and DAP fertilizers are soluble in water.

NPK Fertilizers

NPK fertilizers are three-component fertilizers providing nitrogen, phosphorus, and potassium.

NPK rating is a rating system describing the amount of nitrogen, phosphorus, and potassium in a fertilizer. NPK ratings consist of three numbers separated by dashes (e.g., 10-10-10 or 16-4-8) describing the chemical content of fertilizers. The first number represents the percentage of nitrogen in the product; the second number, P_2O_5; the third, K_2O. Fertilizers do not actually contain P_2O_5 or K_2O, but the system is a conventional shorthand for the amount of the phosphorus (P) or potassium (K) in a fertilizer. A 50-pound (23 kg) bag of fertilizer labeled 16-4-8 contains 8 lb (3.6 kg) of nitrogen (16% of the 50 pounds), an amount of phosphorus equivalent to that in 2 pounds of P_2O_5 (4% of 50 pounds), and 4 pounds of K_2O (8% of 50 pounds). Most fertilizers are labeled according to this N-P-K convention, although Australian convention, following an N-P-K-S system, adds a fourth number for sulfur.

Micronutrients

The main micronutrients are molybdenum, zinc, and copper. These elements are provided as water-soluble salts. Iron presents special problems because it converts to insoluble (bio-unavailable) compounds at moderate soil pH and phosphate concentrations. For this reason, iron is often administered as a chelate complex, e.g., the EDTA derivative. The micronutrient needs depend on the plant. For example, sugar beets appear to require boron, and legumes require cobalt.

Production

Nitrogen Fertilizers

Top users of nitrogen-based fertilizer		
Country	Total N use (Mt pa)	Amt. used for feed/pasture (Mt pa)
China	18.7	3.0
India	11.9	N/A
U.S.	9.1	4.7
France	2.5	1.3
Germany	2.0	1.2
Brazil	1.7	0.7
Canada	1.6	0.9
Turkey	1.5	0.3
UK	1.3	0.9
Mexico	1.3	0.3
Spain	1.2	0.5
Argentina	0.4	0.1

Nitrogen fertilizers are made from ammonia (NH_3), which is sometimes injected into the ground directly. The ammonia is produced by the Haber-Bosch process. In this energy-intensive process, natural gas (CH_4) supplies the hydrogen, and the nitrogen (N_2) is derived from the air. This ammonia is used as a feedstock for all other nitrogen fertilizers, such as anhydrous ammonium nitrate (NH_4NO_3) and urea ($CO(NH_2)_2$).

Deposits of sodium nitrate ($NaNO_3$) (Chilean saltpeter) are also found in the Atacama desert in Chile and was one of the original (1830) nitrogen-rich fertilizers used. It is still mined for fertilizer.

There has been technical work investigating on-site (on-farm) synthesis of nitrate fertilizer using solar photovoltaic power, which would enable farmers more control in soil fertility, while using far less surface area than conventional organic farming for nitrogen fertilizer.

Phosphate Fertilizers

All phosphate fertilizers are obtained by extraction from minerals containing the anion PO_4^{3-}. In rare cases, fields are treated with the crushed mineral, but most often more soluble salts are produced by chemical treatment of phosphate minerals. The most popular phosphate-containing minerals are referred to collectively as phosphate rock. The main minerals are fluorapatite $Ca_5(PO_4)_3F$ (CFA) and hydroxyapatite $Ca_5(PO_4)_3OH$. These minerals are converted to water-soluble phosphate salts by treatment with sulfuric or phosphoric acids. The large production of sulfuric acid as an industrial chemical is primarily due to its use as cheap acid in processing phosphate rock into phosphate fertilizer. The global primary uses for both sulfur and phosphorus compounds relate to this basic process.

In the nitrophosphate process or Odda process (invented in 1927), phosphate rock with up to a 20% phosphorus (P) content is dissolved with nitric acid (HNO_3) to produce a mixture of phosphoric acid (H_3PO_4) and calcium nitrate ($Ca(NO_3)_2$). This mixture can be combined with a potassium fertilizer to produce a *compound fertilizer* with the three macronutrients N, P and K in easily dissolved form.

Potassium Fertilizers

Potash is a mixture of potassium minerals used to make potassium (chemical symbol: K) fertilizers. Potash is soluble in water, so the main effort in producing this nutrient from the ore involves some purification steps; e.g., to remove sodium chloride (NaCl) (common salt). Sometimes potash is referred to as K_2O, as a matter of convenience to those describing the potassium content. In fact potash fertilizers are usually potassium chloride, potassium sulfate, potassium carbonate, or potassium nitrate.

Compound Fertilizers

Compound fertilizers, which contain N, P, and K, can often be produced by mixing straight fertilizers. In some cases, chemical reactions occur between the two or more components. For example, monoammonium and diammonium phosphates, which provide plants with both N and P, are produced by neutralizing phosphoric acid (from phosphate rock) and ammonia :

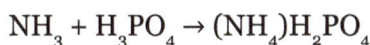

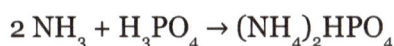

$$NH_3 + H_3PO_4 \rightarrow (NH_4)H_2PO_4$$

$$2\,NH_3 + H_3PO_4 \rightarrow (NH_4)_2HPO_4$$

Organic Fertilizers

Compost bin for small-scale production of organic fertilizer

A large commercial compost operation

The main "organic fertilizers" are peat, animal wastes, plant wastes from agriculture, and treated sewage sludge (biosolids). In terms of volume, peat is the most widely used organic fertilizer. This immature form of coal confers no nutritional value to the plants, but improves the soil by aeration and absorbing water. Animal sources include the products of the slaughter of animals. Bloodmeal, bone meal, hides, hoofs, and horns are typical components. Organic fertilizer usually contain fewer nutrients, but offer other advantages as well as being appealing to those who are trying to practice "environmentally friendly" farming.

Other Elements: Calcium, Magnesium, and Sulfur

Calcium is supplied as superphosphate or calcium ammonium nitrate solutions.

Application

Fertilizers are commonly used for growing all crops, with application rates depending on the soil fertility, usually as measured by a soil test and according to the particular crop. Legumes, for example, fix nitrogen from the atmosphere and generally do not require nitrogen fertilizer.

Liquid vs Solid

Fertilizers are applied to crops both as solids and as liquid. About 90% of fertilizers are applied as solids. Solid fertilizer is typically granulated or powdered. Often solids are available as prills, a solid globule. Liquid fertilizers comprise anhydrous ammonia, aqueous solutions of ammonia, aqueous solutions of ammonium nitrate or urea. These concentrated products may be diluted with water to form a concentrated liquid fertilizer (e.g., UAN). Advantages of liquid fertilizer are its more rapid effect and easier coverage. The addition of fertilizer to irrigation water is called "fertigation".

Slow- and Controlled-release Fertilizers

Slow- and controlled-release involve only 0.15% (562,000 tons) of the fertilizer market (1995). Their utility stems from the fact that fertilizers are subject to antagonistic processes. In addition to their providing the nutrition to plants, excess fertilizers can be poisonous to the same plant. Competitive with the uptake by plants is the degradation or loss of the fertilizer. Microbes degrade many fertilizers, e.g., by immobilization or oxidation. Furthermore, fertilizers are lost by evaporation or leaching. Most slow-release fertilizers are derivatives of urea, a straight fertilizer providing nitrogen. Isobutylidenediurea ("IBDU") and urea-formaldehyde slowly convert in the soil to free urea, which is rapidly uptaken by plants. IBDU is a single compound with the formula $(CH_3)_2CH$-$CH(NHC(O)NH_2)_2$ whereas the urea-formaldehydes consist of mixtures of the approximate formula $(HOCH_2NHC(O)NH)_nCH_2$.

Besides being more efficient in the utilization of the applied nutrients, slow-release technologies also reduce the impact on the environment and the contamination of the subsurface water. Slow-release fertilizers (various forms including fertilizer spikes, tabs, etc.) which reduce the problem of "burning" the plants due to excess nitrogen. Polymer coating of fertilizer ingredients gives tablets and spikes a 'true time-release' or 'staged nutrient release' (SNR) of fertilizer nutrients.

Controlled release fertilizers are traditional fertilizers encapsulated in a shell that degrades at a specified rate. Sulfur is a typical encapsulation material. Other coated products use thermoplastics (and sometimes ethylene-vinyl acetate and surfactants, etc.) to produce diffusion-controlled release of urea or other fertilizers. "Reactive Layer Coating" can produce thinner, hence cheaper, membrane coatings by applying reactive monomers simultaneously to the soluble particles. "Multicote" is a process applying layers of low-cost fatty acid salts with a paraffin topcoat.

Foliar Application

Foliar fertilizers are applied directly to leaves. The method is almost invariably used to apply water-soluble straight nitrogen fertilizers and used especially for high value crops such as fruits.

Fertilizer burn

Chemicals that Affect Nitrogen Uptake

Various chemicals are used to enhance the efficiency of nitrogen-based fertilizers. In this way farmers can limit the polluting effects of nitrogen run-off. Nitrification inhibitors (also known as nitrogen stabilizers) suppress the conversion of ammonia into nitrate, an anion that is more prone to leaching. 1-Carbamoyl-3-methylpyrazole (CMP), dicyandiamide, and nitrapyrin (2-chloro-6-trichloromethylpyridine) are popular. Urease inhibitors are used to slow the hydrolytic conversion of urea into ammonia, which is prone to evaporation as well as nitrification. The conversion of urea to ammonia catalyzed by enzymes called ureases. A popular inhibitor of ureases is N-(n-butyl) thiophosphoric triamide (NBPT).

Overfertilization

Careful fertilization technologies are important because excess nutrients can be as detrimental. Fertilizer burn can occur when too much fertilizer is applied, resulting in drying out of the leaves and damage or even death of the plant. Fertilizers vary in their tendency to burn roughly in accordance with their salt index.

Statistics

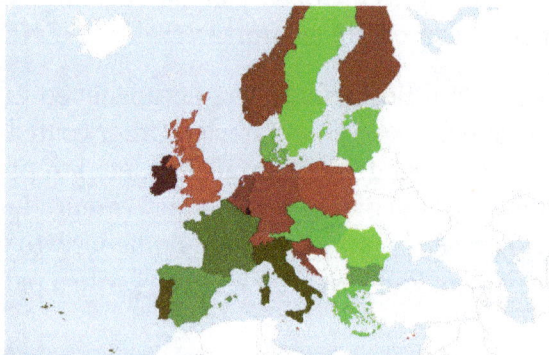

The map displays the statistics of fertilizer consumption in western and central European counties from data published by The World Bank for 2012.

Conservative estimates report 30 to 50% of crop yields are attributed to natural or synthetic commercial fertilizer. Global market value is likely to rise to more than US$185 billion until 2019. The European fertilizer market will grow to earn revenues of approx. €15.3 billion in 2018.

Data on the fertilizer consumption per hectare arable land in 2012 are published by The World Bank. For the diagram below values of the European Union (EU) countries have been extracted and are presented as kilograms per hectare (pounds per acre). The total consumption of fertilizer in the EU is 15.9 million tons for 105 million hectare arable land area (or 107 million hectare arable land according to another estimate). This figure equates to 151 kg of fertilizers consumed per ha arable land on average for the EU countries. Interestingly, mainly in those countries where fertilizers are consumed a lot also plant growth product are sold more than in others.

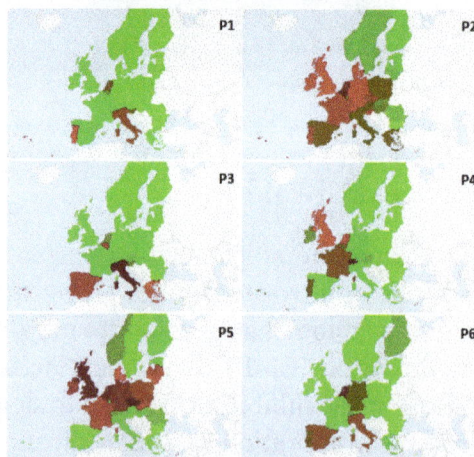

Pesticide categories, EUROSTAT. P5= Plant growth regulators. The red/green scale represents high/low pesticide sales per arable land.

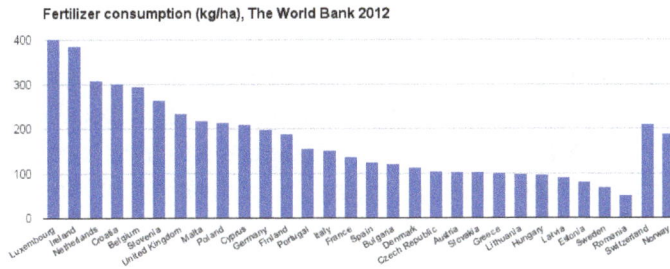
Fertilizer consumption (kg/ha), The World Bank 2012

Environmental Effects

Runoff of soil and fertilizer during a rain storm

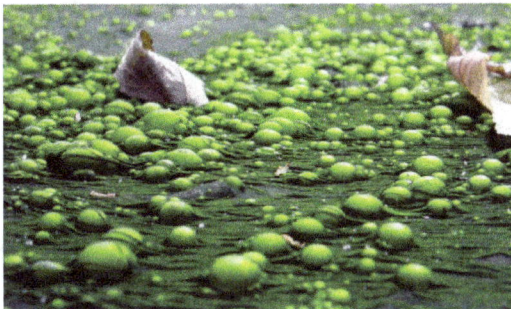
An algal bloom caused by eutrophication

Water

Agricultural run-off is a major contributor to the eutrophication of fresh water bodies. For example, in the US, about half of all the lakes are eutrophic. The main contributor to eutrophication is phosphate, which is normally a limiting nutrient; high concentrations promote the growth of cyanobacteria and algae, the demise of which consumes oxygen. Cyanobacteria blooms ('algal blooms') can also produce harmful toxins that can accumulate in the food chain, and can be harmful to humans.

The nitrogen-rich compounds found in fertilizer runoff are the primary cause of serious oxygen depletion in many parts of oceans, especially in coastal zones, lakes and rivers. The resulting lack of dissolved oxygen greatly reduces the ability of these areas to sustain oceanic fauna. The number of oceanic dead zones near inhabited coastlines are increasing. As of 2006, the application of nitrogen fertilizer is being increasingly controlled in northwestern Europe and the United States. If eutrophication *can* be reversed, it may take decades before the accumulated nitrates in groundwater can be broken down by natural processes.

Nitrate Pollution

Only a fraction of the nitrogen-based fertilizers is converted to produce and other plant matter. The remainder accumulates in the soil or lost as run-off. High application rates of nitrogen-containing fertilizers combined with the high water solubility of nitrate leads to increased runoff into surface water as well as leaching into groundwater, thereby causing groundwater pollution. The excessive use of nitrogen-containing fertilizers (be they synthetic or natural) is particularly damaging, as much of the nitrogen that is not taken up by plants is transformed into nitrate which is easily leached.

Nitrate levels above 10 mg/L (10 ppm) in groundwater can cause 'blue baby syndrome' (acquired methemoglobinemia). The nutrients, especially nitrates, in fertilizers can cause problems for natural habitats and for human health if they are washed off soil into watercourses or leached through soil into groundwater.

Soil

Acidification

Nitrogen-containing fertilizers can cause soil acidification when added. This may lead to decreases in nutrient availability which may be offset by liming.

Accumulation of Toxic Elements

Cadmium

The concentration of cadmium in phosphorus-containing fertilizers varies considerably and can be problematic. For example, mono-ammonium phosphate fertilizer may have a cadmium content of as low as 0.14 mg/kg or as high as 50.9 mg/kg. This is because the phosphate rock used in their manufacture can contain as much as 188 mg/kg cadmium (examples are deposits on Nauru and the Christmas islands). Continuous use of high-cadmium fertilizer can contaminate soil (as shown in New Zealand) and plants. Limits to the cadmium content of phosphate fertilizers has been considered by the European Commission. Producers of phosphorus-containing fertilizers now select phosphate rock based on the cadmium content.

Fluoride

Phosphate rocks contain high levels of fluoride. Consequently, the widespread use of phosphate fertilizers has increased soil fluoride concentrations. It has been found that food contamination from fertilizer is of little concern as plants accumulate little fluoride from the soil; of greater concern is the possibility of fluoride toxicity to livestock that ingest contaminated soils. Also of possible concern are the effects of fluoride on soil microorganisms.

Radioactive Elements

The radioactive content of the fertilizers varies considerably and depends both on their concentrations in the parent mineral and on the fertilizer production process. Uranium-238 concentrations range can range from 7 to 100 pCi/g in phosphate rock and from 1 to 67 pCi/g in phosphate fertilizers. Where high annual rates of phosphorus fertilizer are used, this can result in uranium-238 concentrations in soils and drainage waters that are several times greater than are normally present. However, the impact of these increases on the risk to human health from radinuclide contamination of foods is very small (less than 0.05 mSv/y).

Other Metals

Steel industry wastes, recycled into fertilizers for their high levels of zinc (essential to plant growth), wastes can include the following toxic metals: lead arsenic, cadmium, chromium, and nickel. The most common toxic elements in this type of fertilizer are mercury, lead, and arsenic. These potentially harmful impurities can be removed; however, this significantly increases cost. Highly pure fertilizers are widely available and perhaps best known as the highly water-soluble fertilizers containing blue dyes used around households, such as Miracle-Gro. These highly water-soluble fertilizers are used in the plant nursery business and are available in larger packages at significantly less cost than retail quantities. There are also some inexpensive retail granular garden fertilizers made with high purity ingredients.

Trace Mineral Depletion

Attention has been addressed to the decreasing concentrations of elements such as iron, zinc, copper and magnesium in many foods over the last 50–60 years. Intensive farming practices, including the use of synthetic fertilizers are frequently suggested as reasons for these declines and organic farming is often suggested as a solution. Although improved crop yields resulting from NPK fertilizers are known to dilute the concentrations of other nutrients in plants, much of the measured decline can be attributed to the use of progressively higher-yielding crop varieties which produce foods with lower mineral concentrations than their less productive ancestors. It is, therefore, unlikely that organic farming or reduced use of fertilizers will solve the problem; foods with high nutrient density are posited to be achieved using older, lower-yielding varieties or the development of new high-yield, nutrient-dense varieties.

Fertilizers are, in fact, more likely to solve trace mineral deficiency problems than cause them: In Western Australia deficiencies of zinc, copper, manganese, iron and molybdenum were identified as limiting the growth of broad-acre crops and pastures in the 1940s and 1950s. Soils in Western Australia are very old, highly weathered and deficient in many of the major nutrients and trace elements. Since this time these trace elements are routinely added to fertilizers used in agriculture in this state. Many other soils around the world are deficient in zinc, leading to deficiency in both plants and humans, and zinc fertilizers are widely used to solve this problem.

Changes in Soil Biology

High levels of fertilizer may cause the breakdown of the symbiotic relationships between plant roots and mycorrhizal fungi.

Energy Consumption and Sustainability

In the USA in 2004, 317 billion cubic feet of natural gas were consumed in the industrial production of ammonia, less than 1.5% of total U.S. annual consumption of natural gas. A 2002 report suggested that the production of ammonia consumes about 5% of global natural gas consumption, which is somewhat under 2% of world energy production.

Ammonia is produced from natural gas and air. The cost of natural gas makes up about 90% of the cost of producing ammonia. The increase in price of natural gases over the past decade, along with other factors such as increasing demand, have contributed to an increase in fertilizer price.

Contribution to Climate Change

The greenhouse gases carbon dioxide, methane and nitrous oxide are produced during the manufacture of nitrogen fertilizer. The effects can be combined into an equivalent amount of carbon dioxide. The amount varies according to the efficiency of the process. The figure for the United Kingdom is over 2 kilogrammes of carbon dioxide equivalent for each kilogramme of ammonium nitrate. Nitrogen fertilizer can be converted by soil bacteria to nitrous oxide, a greenhouse gas.

Atmosphere

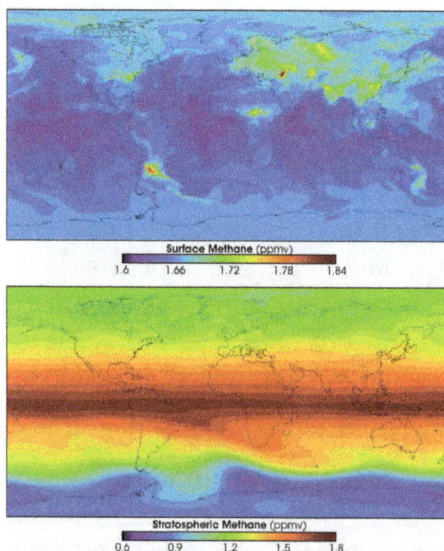

Global methane concentrations (surface and atmospheric) for 2005; note distinct plumes

Through the increasing use of nitrogen fertilizer, which was used at a rate of about 110 million tons (of N) per year in 2012, adding to the already existing amount of reactive nitrogen, nitrous oxide (N_2O) has become the third most important greenhouse gas after carbon dioxide and methane. It has a global warming potential 296 times larger than an equal mass of carbon dioxide and it also contributes to stratospheric ozone depletion. By changing processes and procedures, it is possible to mitigate some, but not all, of these effects on anthropogenic climate change.

Methane emissions from crop fields (notably rice paddy fields) are increased by the application of ammonium-based fertilizers. These emissions contribute to global climate change as methane is a potent greenhouse gas.

Regulation

In Europe problems with high nitrate concentrations in run-off are being addressed by the European Union's Nitrates Directive. Within Britain, farmers are encouraged to manage their land more sustainably in 'catchment-sensitive farming'. In the US, high concentrations of nitrate and phosphorus in runoff and drainage water are classified as non-point source pollutants due to their diffuse origin; this pollution is regulated at state level. Oregon and Washington, both in the United States, have fertilizer registration programs with on-line databases listing chemical analyses of fertilizers.

History

Founded in 1812, Mirat, producer of manures and fertilizers, is claimed to be the oldest industrial business in Salamanca (Spain).

Management of soil fertility has been the preoccupation of farmers for thousands of years. Egyptians, Romans, Babylonians, and early Germans all are recorded as using minerals and or manure to enhance the productivity of their farms. The modern science of plant nutrition started in the 19th century and the work of German chemist Justus von Liebig, among others. John Bennet Lawes, an English entrepreneur, began to experiment on the effects of various manures on plants growing in pots in 1837, and a year or two later the experiments were extended to crops in the field. One immediate consequence was that in 1842 he patented a manure formed by treating phosphates with sulphuric acid, and thus was the first to create the artificial manure industry. In the succeeding year he enlisted the services of Joseph Henry Gilbert, with whom he carried on for more than half a century on experiments in raising crops at the Institute of Arable Crops Research.

The Birkeland–Eyde process was one of the competing industrial processes in the beginning of nitrogen based fertilizer production. This process was used to fix atmospheric nitrogen (N_2) into nitric acid (HNO_3), one of several chemical processes generally referred to as nitrogen fixation. The resultant nitric acid was then used as a source of nitrate (NO_3^-). A factory based on the process was built in Rjukan and Notodden in Norway, combined with the building of large hydroelectric power facilities.

The 1910s and 1920s witness the rise of the Haber process and the Ostwald process. The Haber process produces ammonia (NH_3) from methane (CH_4) gas and molecular nitrogen (N_2). The ammonia from the Haber process is then converted into nitric acid (HNO_3) in the Ostwald process. The development of synthetic fertilizer has significantly supported global population growth — it has been estimated that almost half the people on the Earth are currently fed as a result of synthetic nitrogen fertilizer use.

The use of commercial fertilizers has increased steadily in the last 50 years, rising almost 20-fold to the current rate of 100 million tonnes of nitrogen per year. Without commercial fertilizers it is estimated that about one-third of the food produced now could not be produced. The use of phosphate fertilizers has also increased from 9 million tonnes per year in 1960 to 40 million tonnes per year in 2000. A maize crop yielding 6–9 tonnes of grain per hectare (2.5 acres) requires 31–50 kilograms (68–110 lb) of phosphate fertilizer to be applied; soybean crops require about half, as 20–25 kg per hectare. Yara International is the world's largest producer of nitrogen-based fertilizers.

Controlled-nitrogen-release technologies based on polymers derived from combining urea and formaldehyde were first produced in 1936 and commercialized in 1955. The early product had 60 percent of the total nitrogen cold-water-insoluble, and the unreacted (quick-release) less than 15%. Methylene ureas were commercialized in the 1960s and 1970s, having 25% and 60% of the nitrogen as cold-water-insoluble, and unreacted urea nitrogen in the range of 15% to 30%.

In the 1960s, the Tennessee Valley Authority National Fertilizer Development Center began developing sulfur-coated urea; sulfur was used as the principal coating material because of its low cost and its value as a secondary nutrient. Usually there is another wax or polymer which seals the sulfur; the slow-release properties depend on the degradation of the secondary sealant by soil microbes as well as mechanical imperfections (cracks, etc.) in the sulfur. They typically provide 6 to 16 weeks of delayed release in turf applications. When a hard polymer is used as the secondary coating, the properties are a cross between diffusion-controlled particles and traditional sulfur-coated.

Soil pH

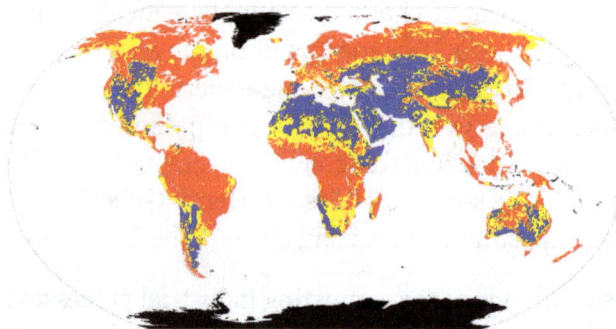

Global variation in soil pH. Red = acidic soil. Yellow = neutral soil. Blue = alkaline soil. Black = no data.

The soil pH is a measure of the acidity or alkalinity in soils. pH is defined as the negative logarithm (base 10) of the activity of hydronium ions (H+or, more precisely, H3O+aq) in a solution. In water, it normally ranges from -1 to 14, with 7 being neutral. A pH below 7 is acidic and above 7 is alkaline. Soil pH is considered a master variable in soils as it controls many chemical processes that take place. It specifically affects plant nutrient availability by controlling the chemical forms of the nutrient. The optimum pH range for most plants is between 5.5 and 7.0, however many plants have adapted to thrive at pH values outside this range.

Classification of Soil pH Ranges

The United States Department of Agriculture Natural Resources Conservation Service, formerly

Soil Conservation Service classifies soil pH ranges as follows:

Denomination	pH range
Ultra acidic	< 3.5
Extremely acidic	3.5–4.4
Very strongly acidic	4.5–5.0
Strongly acidic	5.1–5.5
Moderately acidic	5.6–6.0
Slightly acidic	6.1–6.5
Neutral	6.6–7.3
Slightly alkaline	7.4–7.8
Moderately alkaline	7.9–8.4
Strongly alkaline	8.5–9.0
Very strongly alkaline	> 9.0

Sources of Soil pH

Sources of Acidity

Acidity in soils comes from H^+ and Al^{3+} ions in the soil solution and sorbed to soil surfaces. While pH is the measure of H^+ in solution, Al^{3+} is important in acid soils because between pH 4 and 6, Al^{3+} reacts with water (H_2O) forming $AlOH^{2+}$, and $Al(OH)_2^+$, releasing extra H^+ ions. Every Al^{3+} ion can create 3 H^+ ions. Many other processes contribute to the formation of acid soils including rainfall, fertilizer use, plant root activity and the weathering of primary and secondary soil minerals. Acid soils can also be caused by pollutants such as acid rain and mine spoilings.

- Rainfall: Acid soils are most often found in areas of high rainfall. Excess rainfall leaches base cation from the soil, increasing the percentage of Al^{3+} and H^+ relative to other cations. Additionally, rainwater has a slightly acidic pH of 5.7 due to a reaction with CO_2 in the atmosphere that forms carbonic acid.

- Fertilizer use: Ammonium (NH_4^+) fertilizers react in the soil in a process called nitrification to form nitrate (NO_3^-), and in the process release H^+ ions.

- Plant root activity: Plants take up nutrients in the form of ions (NO_3^-, NH_4^+, Ca^{2+}, $H_2PO_4^-$, etc.), and often, they take up more cations than anions. However plants must maintain a neutral charge in their roots. In order to compensate for the extra positive charge, they will release H^+ ions from the root. Some plants will also exude organic acids into the soil to acidify the zone around their roots to help solubilize metal nutrients that are insoluble at neutral pH, such as iron (Fe).

- Weathering of minerals: Both primary and secondary minerals that compose soil contain Al. As these minerals weather, some components such as Mg, Ca, and K, are taken up by plants, others such as Si are leached from the soil, but due to chemical properties, Fe and

Al remain in the soil profile. Highly weathered soils are often characterized by having high concentrations of Fe and Al oxides.

- Acid rain: When atmospheric water reacts with sulfur and nitrogen compounds that result from industrial processes, the result can be the formation of sulfuric and nitric acid in rainwater. However the amount of acidity that is deposited in rainwater is much less, on average, than that created through agricultural activities.

- Mine spoil: Severely acidic conditions can form in soils near mine spoils due to the oxidation of pyrite.

- Potential acid sulfate soils naturally formed in waterlogged coastal and estuarine environments can become highly acidic when drained or excavated.

- Decomposition of organic matter by microorganisms releases CO_2 which when mixed with soil water can form carbonic acid (H_2CO_3).

Sources of Alkalinity

Alkaline soils have a high saturation of base cations (K^+, Ca^{2+}, Mg^{2+} and Na^+). This is due to an accumulation of soluble salts which are classified as either saline soil, sodic soil, saline-sodic soil or alkaline soil. All saline and sodic soils have high salt concentrations, with saline soils being dominated by calcium and magnesium salts and sodic soils being dominated by sodium. Alkaline soils are characterized by the presence of carbonates. Soil in areas with limestone near the surface are alkaline from the calcium carbonate in limestone constantly mixing with the soil. Groundwater sources in these areas contain dissolved limestone.

Effect of Soil pH on Plant Growth

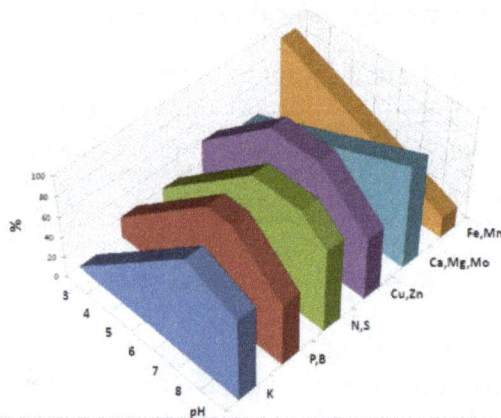

Nutrient availability in relation to soil pH

Acid Affected Soils

Plants grown in acid soils can experience a variety of symptoms including aluminium (Al), hydrogen (H), and/or manganese (Mn) toxicity, as well as nutrient deficiencies of calcium (Ca) and magnesium (Mg).

Aluminium toxicity is the most widespread problem in acid soils. Aluminium is present in all soils, but dissolved Al^{3+} is toxic to plants; Al^{3+} is most soluble at low pH, above pH 5.2 little Al is in soluble form in most soils. Aluminium is not a plant nutrient, and as such, is not actively taken up by the plants, but enters plant roots passively through osmosis. Aluminium inhibits root growth; lateral roots and root tips become thickened and roots lack fine branching; root tips may turn brown. In the root, Al has been shown to interfere with many physiological processes including the uptake and transport of calcium and other essential nutrients, cell division, cell wall formation, and enzyme activity.

Below pH 4, H^+ ions themselves damage root cell membranes.

In soils with high content of manganese-containing minerals, Mn toxicity can become a problem at pH 5.6 and lower. Manganese, like aluminium, becomes increasingly soluble as pH drops, and Mn toxicity symptoms can be seen at pH levels below 5.6. Manganese is an essential plant nutrient, so plants transport Mn into leaves. Classic symptoms of Mn toxicity are crinkling or cupping of leaves.

Nutrient Availability in Relation to Soil pH

Nutrients needed in large amounts by plants are referred to as macronutrients and include nitrogen (N), phosphorus (P), potassium (K), calcium (Ca), magnesium (Mg) and sulfur (S). Elements that plants need in trace amounts are called trace nutrients or micronutrients. Trace nutrients are not major components of plant tissue but are essential for growth. They include iron (Fe), manganese (Mn), zinc (Zn), copper (Cu), cobalt (Co), molybdenum (Mo), and boron (B). Both macronutrient and micronutrient availability are affected by soil pH. In slightly to moderately alkaline soils, molybdenum and macronutrient (except for phosphorus) availability is increased, but P, Fe, Mn, Zn Cu, and Co levels are reduced and may adversely affect plant growth. In acidic soils, micronutrient availability (except for Mo and Bo) is increased. Nitrogen is supplied as ammonium (NH4) or nitrate (NO3) by nitrogen fixation or fertilizer amendments, and dissolved N will have the highest concentrations in soil with pH 6–8. Concentrations of available N are less sensitive to pH than concentration of available P. In order for P to be available for plants, soil pH needs to be in the range 6.0 and 7.5. If pH is lower than 6, P starts forming insoluble compounds with iron (Fe) and aluminium (Al) and if pH is higher than 7.5 P starts forming insoluble compounds with calcium (Ca). Most nutrient deficiencies can be avoided between a pH range of 5.5 to 6.5, provided that soil minerals and organic matter contain the essential nutrients to begin with.

Water Availability in Relation to Soil pH

Determining pH

Methods of determining pH include:

- Observation of soil profile: Certain profile characteristics can be indicators of either acid, saline, or sodic conditions. Strongly acidic soils often have poor incorporation of the organic surface layer with the underlying mineral layer. The mineral horizons are distinctively layered in many cases, with a pale eluvial (E) horizon beneath the organic surface; this E is underlain by a darker B horizon in a classic podzol horizon sequence. This is a very rough gauge of acidity as there is no correlation between thickness of the E and soil pH. E horizons a few feet thick in Florida usually have pH just above 5 (merely "strongly acid") while E horizons a few inches thick in New England are "extremely acid" with pH readings of 4.5 or below. In the southern Blue Ridge Mountains there are "ultra acid" soils, pH below 3.5, which have no E horizon. Presence of a caliche layer indicates the presence of calcium carbonates, which are present in alkaline conditions. Also, columnar structure can be an indicator of sodic condition.

- Observation of predominant flora. Calcifuge plants (those that prefer an acidic soil) include *Erica*, *Rhododendron* and nearly all other Ericaceae species, many birch (*Betula*), foxglove (*Digitalis*), gorse (*Ulex* spp.), and Scots Pine (*Pinus sylvestris*). Calcicole (lime loving) plants include ash trees (*Fraxinus* spp.), honeysuckle (*Lonicera*), *Buddleja*, dogwoods (*Cornus* spp.), lilac (*Syringa*) and *Clematis* species.

- Use of an inexpensive pH testing kit, where in a small sample of soil is mixed with indicator solution which changes colour according to the acidity/alkalinity.

- Use of litmus paper. A small sample of soil is mixed with distilled water, into which a strip of litmus paper is inserted. If the soil is acidic the paper turns red, if alkaline, blue.

- Use of a commercially available electronic pH meter, in which a rod is inserted into moistened soil and measures the concentration of hydrogen ions.

Examples of Plant pH Preferences

- pH 4.5–5.0: Ericaceae (azalea, bilberry, blueberry, cranberry; heather), hydrangea for blue, (less acidic for pink), liquidambar or sweet gum, orchid, pin oak

- pH 5.0–5.5: Boronia, daphne; Ericaceae: (camellia, heather, rhododendron), ferns, iris, orchids, parsley, conifers (e.g., pine); Poaceae: (maize, millet, rye, oat), radish; Solanales: (potato, sweet potato); Bromeliaceae (pineapple).

- pH 5.5–6.0: Asteraceae: (aster, endive); Brassicaceae: (brussels sprout, kohlrabi), carrot; Cucurbitales: (begonia, chayote or choko); Fabaceae: (bean, crimson clover, peanut, soybean), petunia, rhubarb, violet, most bulbs (canna, daffodil, jonquil), larkspur, primrose

- pH 6.0–6.5 antirrhinum or snapdragon, Brassicaceae: (broccoli, cabbage, candytuft, cauliflower, turnip, wallflower); Cucurbitaceae: (cucumber, pumpkin, squash); Fabaceae: (pea, red clover, white clover), gladiolus, Iceland poppy; Rosales: (cannabis, rose, strawberry); Solanaceae: (eggplant or aubergine, tomato), sweet corn; Violaceae: (pansy, viola), zinnia or zinnea

- pH 6.5–7.0: Amaranthaceae: (beet, spinach); Apiaceae: (celery, parsnip); Asparagales: (asparagus, onion); Asteraceae: (chrysanthemum, dahlia, lettuce), carnation; Fabaceae: (alfalfa, sweet pea), melon, stock, tulip

- pH 7.1–8.0 lilac

Changing Soil pH

Increasing pH of Acidic Soil

The most common amendment to increase soil pH is lime ($CaCO_3$ or $MgCO_3$), usually in the form of finely ground agricultural lime. The amount of lime needed to change pH is determined by the mesh size of the lime (how finely it is ground)and the buffering capacity of the soil. A high mesh size (60–100) indicates a finely ground lime, that will react quickly with soil acidity. Buffering capacity of soils is a function of a soils cation exchange capacity, which is in turn determined by the clay content of the soil, the type of clay and the amount of organic matter present. Soils with high clay content, particularly shrink–swell clay, will have a higher buffering capacity than soils with little clay. Soils with high organic matter will also have a higher buffering capacity than those with low organic matter. Soils with high buffering capacity require a greater amount of lime to be added than a soil with a lower buffering capacity for the same incremental change in pH. Other amendments that can be used to increase the pH of soil include wood ash, industrial CaO (burnt lime), and oyster shells. White firewood ash includes metal salts which are important for processes requiring ions such as Na^+ (sodium), K^+ (potassium), Ca^{2+} (calcium), which may or may not be good for the select flora, but decreases the acidic quality of soil. These products increase the pH of soils through the reaction of CO_3^{2-} with H^+ to produce CO_2 and H_2O. Calcium silicate neutralizes active acidity in the soil by removing free hydrogen ions, thereby increasing pH. As its silicate anion captures H^+ ions (raising the pH), it forms monosilicic acid (H_4SiO_4), a neutral solute.

Decreasing pH of Alkaline Soil

- Iron sulfates or aluminium sulfate as well as elemental sulfur (S) reduce pH through the formation of sulfuric acid.

- Urea, urea phosphate, ammonium nitrate, ammonium phosphates, ammonium sulfate and monopotassium phosphate fertilizers.

- organic matter in the form of plant litter, compost, and manure will decrease soil pH through the decomposition process. Certain acid organic matter such as pine needles, pine sawdust and acid peat are effective at reducing pH.

Cation-exchange Capacity

In soil science, cation-exchange capacity or CEC is the number of exchangeable cations per dry weight that a soil is capable of holding, at a given pH value, and available for exchange with the soil water solution. CEC is used as a measure of soil fertility, nutrient retention capacity, and the capacity to protect groundwater from cation contamination. It is expressed as milliequivalent of hydrogen per 100 g of dry soil (meq+/100g), or the SI unit centi-mol per kg (cmol+/kg). The numeric values are the same in any system of units.

Clay and humus have electrostatic surface charges that attract and hold ions. The holding capacity of clay varies with the type of clay. Humus has a CEC that is two to three times that of the best clay.

One way to increase the CEC of a soil is to enhance the formation of humus.

In general, the higher the CEC, the higher the fertility of that soil.

Calculation of CEC

The CEC is the number of positive charges (cations) that a representative sample of soil can hold. It is usually described as the number of hydrogen ions (H^+) necessary to fill the soil cation holding sites per 100 grams of dry soil. Alternatively, an equivalent amount of another cation (Al^{3+} or Ca^{2+}) can be used in the measure. In soil science, an equivalent is defined by the number of charges in terms of a given number of hydrogen ions. As hydrogen ions have only one positive charge (H^+), this makes calculations relatively simple. An equivalent of Al^{3+} that could be held would amount to one third as many of those ions, and Ca^{2+} would have half as many ions.

Translation from meq/100g to an applicable unit, like lb/acre of available nutrients, can be made via calculation, that considers the atomic weight, the ion's valence, and by estimating the soil depth and its density. Mengel gives the following correspondence for 1 meq/100g:

> Calcium, 400 lb/acre
>
> Magnesium, 240 lb/acre
>
> Potassium, 780 lb/acre
>
> Ammonium, 360 lb/acre

Base Saturation

Closely related to cation-exchange capacity is the base saturation, which is the fraction of exchangeable cations that are base cations (Ca, Mg, K and Na). It can be expressed as a percentage, and called *percent base saturation*. The higher the amount of exchangeable base cations, the more acidity can be neutralised in the short time perspective. Thus, a soil with high cation-exchange capacity takes longer time to acidify (as well as to recover from an acidified status) than a soil with a low cation-exchange capacity (assuming similar base saturations). A rain, with its load of acidic hydrogen ions, upon a soil that has a high CEC will be quickly returned (buffered) to its original pH in a very short time. A rain on a low CEC soil, such as in the Amazon Basin with its acid soils, will not be restored and the pH will drop sharply and remain there for a relatively long time.

The base-cation saturation ratio (BCSR) is a method of interpreting soil test results that is widely used in sustainable agriculture, supported by the National Sustainable Agriculture Information Service (ATTRA) and claimed to be successfully in use on over a million acres (4,000 km²) of farmland worldwide.

pH and CEC

For many soils, the CEC is dependent upon the pH of the soil. This is due mostly to the Hofmeister series (lyotrophic series), which describes the relative strength of various cations' adsorption to colloids, and is generally as follows:

$$Al^{3+} > H^+ > Ca^{2+} > Mg^{2+} > K^+ = NH_4^+ > Na^+$$

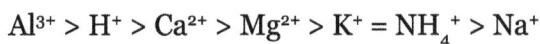

As soil acidity increases (pH decreases), more H^+ ions are attached to the colloids. They have pushed the other cations from the colloids and into the soil water solution. Inversely, when soils become more basic (pH increases), the available cations in solution decreases because there are fewer H^+ ions to push cations into the soil solution from the colloids (CEC increases).

Various Colloids and Soils' CEC

The CEC of various soils and soil constituents varies greatly.

Cation exchange capacity for soils; soil textures; soil colloids		
Soil	**State**	**CEC meq/100 g**
Charlotte fine sand	Florida	1.0
Ruston fine sandy loam	Texas	1.9
Glouchester loam	New Jersey	11.9
Grundy silt loam	Illinois	26.3
Gleason clay loam	California	31.6
Susquehanna clay loam	Alabama	34.3
Davie mucky fine sand	Florida	100.8
Sands	—	1–5
Fine sandy loams	—	5–10
Loams and silt loams	—	5–15
Clay loams	—	15–30
Clays	—	over 30
Sesquioxides	—	0–3
Kaolinite	—	3–15
Illite	—	25–40
Montmorillonite	—	60–100
Vermiculite (similar to illite)	—	80–150
Humus	—	100–300

Aluminium Ions and CEC

Many heavily leached or oxidized soils, especially in the wet tropics, have a high concentration of Al^{3+} occupying the soil colloids cation exchange sites. Since aluminium is toxic in high quantities for most plants, there are certain advantages to this. Due to the relatively high adsorption rate of aluminium to soil colloids, it will be taken out of the soil, hence the plant cannot be adversely affected by it. On the other hand, because it has three positive charges, it takes up a large amount of charge on a colloid. For example, Al^{3+} fills the same space as three NH_4^+ ions. As a result, the ammonium is left in the soil water solution where it can be washed away by a heavy rain. This makes many aluminium heavy soils relatively infertile. There is no easy way to remove aluminium ions from the soil colloid and free the CEC for other ions.

Organic Matter

Organic materials in soil increase the CEC through an increase in available negative charges. As such, organic matter build-up in soil usually positively impacts soil fertility. However, organic matter CEC is heavily impacted by soil acidity as acidity causes many organic compounds to release ions to the soil solution.

Anion Exchange Capacity

Similar to the CEC, the anion exchange capacity is a measurement of the positive charges in soils affecting the amount of negative charges which a soil can absorb. There are relatively few anions that are restrictive in agriculture, but they are important, such as sulfur or phosphorus. The anion lyotrophic series is:

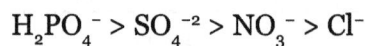

$$H_2PO_4^- > SO_4^{-2} > NO_3^- > Cl^-$$

Converse to CEC, AEC generally will increase when pH drops and decrease when pH rises.

Laboratory Determination

There are two standardised International Soil Reference and Information Centre methods for determining CEC:

- extraction with ammonium acetate; and

- the silver-thiourea method (one-step centrifugal extraction).

There exist slightly conflicting ideas on which mechanisms to include in the term, "cation exchange", in soil chemistry. From a theoretical point of view, one should distinguish cation exchange from ligand exchange, and exchange of diffuse layer adsorbed cations. On the other hand, from a practical point of view, e.g. in forest and agricultural management, what is important is the soils' ability to replace one cation with another rather than the exact mechanism by which this replacement occurs. What is included in the term, "cation exchange", in soil science thus varies with the scientific context.

Standard Values

Kaolinite	3–15
Halloysite $2H_2O$	5–10
Halloysite $4H_2O$	40–50
Montmorillonite-group	70–100
Illite	10–40
Vermiculite	100–150

Chlorite	10–40
Glauconite	11–20+
Palygorskite-group	20–30
Allophane	~70

These are the values reported by Carroll (1959) for the cation-exchange capacity of minerals in meq/100g at pH of 7.

Soil Test

Soil test may refer to one or more of a wide variety of soil analyses conducted for one of several possible reasons. Possibly the most widely conducted soil tests are those done to estimate the plant-available concentrations of plant nutrients, in order to determine fertilizer recommendations in agriculture. Other soil tests may be done for engineering (geotechnical), geochemical or ecological investigations.

Plant Nutrition

In agriculture, a soil test commonly refers to the analysis of a soil sample to determine nutrient content, composition, and other characteristics such as the acidity or pH level. A soil test can determine fertility, or the expected growth potential of the soil which indicates nutrient deficiencies, potential toxicities from excessive fertility and inhibitions from the presence of non-essential trace minerals. The test is used to mimic the function of roots to assimilate minerals. The expected rate of growth is modeled by the Law of the Maximum.

Labs, such as those at Iowa State and Colorado State University, recommend that a soil test contains 10-20 sample points for every 40 acres (160,000 m²) of field. Tap water or chemicals can change the composition of the soil, and may need to be tested separately. As soil nutrients vary with depth and soil components change with time, the depth and timing of a sample may also affect results.

Composite sampling can be performed by combining soil from several locations prior to analysis. This is a common procedure, but should be used judiciously to avoid skewing results. This procedure must be done so that government sampling requirements are met. A reference map should be created to record the location and quantity of field samples in order to properly interpret test results.

Storage, Handling, and Moving

Soil chemistry changes over time, as biological and chemical processes break down or combine compounds over time. These processes change once the soil is removed from its natural ecosystem (flora and fauna that penetrate the sampled area) and environment (temperature, moisture, and solar light/radiation cycles). As a result, the chemical composition analysis accuracy can be improved if the soil is analysed soon after its extraction — usually within a relative time period of

24 hours. The chemical changes in the soil can be slowed during storage and transportation by freezing it. Air drying can also preserve the soil sample for many months.

Soil Testing

Soil testing is often performed by commercial labs that offer a variety of tests, targeting groups of compounds and minerals. The advantages associated with local lab is that they are familiar with the chemistry of the soil in the area where the sample was taken. This enables technicians to recommend the tests that are most likely to reveal useful information.

Soil testing in progress

Laboratory tests often check for plant nutrients in three categories:

- Major nutrients: nitrogen (N), phosphorus (P), and potassium (K)

- Secondary nutrients: sulfur, calcium, magnesium

- Minor nutrients: iron, manganese, copper, zinc, boron, molybdenum, chlorine

Do-it-yourself kits usually only test for the three "major nutrients", and for soil acidity or pH level. Do-it-yourself kits are often sold at farming cooperatives, university labs, private labs, and some hardware and gardening stores. Electrical meters that measure pH, water content, and sometimes nutrient content of the soil are also available at many hardware stores. Laboratory tests are more accurate than tests with do-it-yourself kits and electrical meters. Here is an example soil sample report from one laboratory.

Soil testing is used to facilitate fertilizer composition and dosage selection for land employed in both agricultural and horticultural industries.

Prepaid mail-in kits for soil and ground water testing are available to facilitate the packaging and delivery of samples to a laboratory. Similarly, in 2004, laboratories began providing fertilizer recommendations along with the soil composition report.

Lab tests are more accurate, though both types are useful. In addition, lab tests frequently include professional interpretation of results and recommendations. Always refer to all proviso statements included in a lab report as they may outline any anomalies, exceptions, and shortcomings in the sampling and/or analytical process/results.

Some laboratories analyze for all 13 mineral nutrients and a dozen non-essential, potentially toxic minerals utilizing the "universal soil extractant" (ammonium bicarbonate DTPA).

Soil Contaminants

Common mineral soil contaminants include arsenic, barium, cadmium, copper, mercury, lead, and zinc.

Lead is a particularly dangerous soil component. The following table from the University of Minnesota categorizes typical soil concentration levels and their associated health risks.

Children and pregnant women should avoid contact with soil estimated total lead levels above 300 ppm		
Lead Level	**Extracted lead (ppm)**	**Estimated total lead (ppm)**
Low	<43	<500
Medium	43-126	500-1000
High	126-480	1000-3000
Very high	>480	>3000

Six gardening practices to reduce the lead risk

1. Locate gardens away from old painted structures and heavily traveled roads

2. Give planting preferences to fruiting crops (tomatoes, squash, peas, sunflowers, corn, etc.)

3. Incorporate organic materials such as finished compost, humus, and peat moss

4. Lime soil as recommended by soil test (pH 6.5 minimizes lead availability)

5. Discard old and outer leaves before eating leafy vegetables; peel root crops; wash all produce

6. Keep dust to a minimum by maintaining a mulched and/or moist soil surface

References

- F. Stuart Chapin III; Pamela A. Matson; Harold A. Moon (2002). Principles of Terrestrial Ecosystem Ecology. Springer. ISBN 0387954392.

- Kötke, William H. (1993). The Final Empire: The Collapse of Civilization and the Seed of the Future. Arrow Point Press. ISBN 0963378457.

- Bill Mollison, Permaculture: A Designer's Manual, Tagari Press, (December 1, 1988), 576 pages, ISBN 0908228015. Increases in porosity enhance infiltration and thus reduce adverse effects of surface runoff.

- Arthur T. Hubbard, Encyclopedia of Surface and Colloid Science Vol 3, Santa Barbara, California Science Project, Marcel Dekker, New York (2004) ISBN 0-8247-0759-1

- Moore, Geoff (2001). Soilguide - A handbook for understanding and managing agricultural soils (PDF). Perth, Western Australia: Agriculture Western Australia. pp. 161–207. ISBN 0 7307 0057 7.

- Carroll and Salt, Steven B. and Steven D. (2004). Ecology for Gardeners. Cambridge: Timber Press. ISBN 9780881926118.

- Aaron John Ihde (1984). The development of modern chemistry. Courier Dover Publications. p. 678. ISBN 0-486-64235-6.

- G. J. Leigh (2004). The world's greatest fix: a history of nitrogen and agriculture. Oxford University Press US. pp. 134–139. ISBN 0-19-516582-9.

- Donahue, Miller, Shickluna (1977). Soils: an introduction to soils and plant growth (4 ed.). Inglewood Cliffs, New Jersey 07632: Prentice- Hall. pp. 115, 116. ISBN 0-13-821918-4.

- FAO (2012). Current world fertilizer trends and outlook to 2016 (PDF). Rome: Food and Agriculture Organization of the United Nations. p. 13. Retrieved 3 July 2014.

- "A Farmer's Guide To Agriculture and Water Quality Issues: 3. Environmental Requirements & Incentive Programs For Nutrient Management". www.cals.ncsu.edu. Retrieved 3 July 2014.

Plant Nutrition

Plant nutrition is the analysis of chemical elements that are vital for plant growth and plant metabolism. For plant nutrition, water potential and transpiration are also important processes involved in the growth of a plant. This text helps the reader in understanding the concept of plant nutrition.

Plant Nutrition

Plant nutrition is the study of the chemical elements and compounds necessary for plant growth, plant metabolism and their external supply. In 1972, E. Epstein defined two criteria for an element to be essential for plant growth:

1. in its absence the plant is unable to complete a normal life cycle.

2. or that the element is part of some essential plant constituent or metabolite.

This is in accordance with Justus von Liebig's law of the minimum. The essential plant nutrients include carbon and oxygen which are absorbed from the air, whereas other nutrients including hydrogen are typically obtained from the soil (exceptions include some parasitic or carnivorous plants).

Plants must obtain the following mineral nutrients from their growing medium:

- the macronutrients: nitrogen (N), phosphorus (P), potassium (K), calcium (Ca), sulfur (S), magnesium (Mg); and

- the micronutrients (or trace minerals): boron (B), chlorine (Cl), manganese (Mn), iron (Fe), zinc (Zn), copper (Cu), molybdenum (Mo), nickel (Ni).

Farmer spreading decomposing manure to improve soil fertility and plant nutrition

The macronutrients are consumed in larger quantities and are usually present in plant tissue in concentrations of between 0.2% and 4.0% (on a dry matter weight basis). Micronutrients are present in plant tissue in quantities measured in parts per million, ranging from 0.1 to 200 ppm, or less than 0.02% dry weight.

Most soil conditions across the world can provide plants adapted to that climate and soil with sufficient nutrition for a complete life cycle, without the addition of nutrients as fertilizer. However, if the soil is cropped it is necessary to artificially modify soil fertility through the addition of fertilizer to promote vigorous growth and increase or sustain yield. This is done because, even with adequate water and light, nutrient deficiency can limit growth and crop yield.

Processes

Plants take up essential elements from the soil through their roots and from the air (mainly consisting of nitrogen and oxygen) through their leaves. Nutrient uptake in the soil is achieved by cation exchange, wherein root hairs pump hydrogen ions (H^+) into the soil through proton pumps. These hydrogen ions displace cations attached to negatively charged soil particles so that the cations are available for uptake by the root. In the leaves, stomata open to take in carbon dioxide and expel oxygen. The carbon dioxide molecules are used as the carbon source in photosynthesis.

The root, especially the root hair, is the essential organ for the uptake of nutrients. The structure and architecture of the root can alter the rate of nutrient uptake. Nutrient ions are transported to the center of the root, the stele in order for the nutrients to reach the conducting tissues, xylem and phloem. The Casparian strip, a cell wall outside the stele but within the root, prevents passive flow of water and nutrients, helping to regulate the uptake of nutrients and water. Xylem moves water and inorganic molecules within the plant and phloem accounts for organic molecule transportation. Water potential plays a key role in a plant's nutrient uptake. If the water potential is more negative within the plant than the surrounding soils, the nutrients will move from the region of higher solute concentration—in the soil—to the area of lower solute concentration - in the plant.

There are three fundamental ways plants uptake nutrients through the root:

1. Simple diffusion occurs when a nonpolar molecule, such as O_2, CO_2, and NH_3 follows a concentration gradient, moving passively through the cell lipid bilayer membrane without the use of transport proteins.

2. Facilitated diffusion is the rapid movement of solutes or ions following a concentration gradient, facilitated by transport proteins.

3. Active transport is the uptake by cells of ions or molecules against a concentration gradient; this requires an energy source, usually ATP, to power molecular pumps that move the ions or molecules through the membrane.

Nutrients can be moved within plants to where they are most needed. For example, a plant will try to supply more nutrients to its younger leaves than to its older ones. When nutrients are mobile within the plant, symptoms of any deficiency become apparent first on the older leaves. However, not all nutrients are equally mobile. Nitrogen, phosphorus, and potassium are mobile nutrients while the others have varying degrees of mobility. When a less-mobile nutrient is deficient, the younger leaves suffer because the nutrient does not move up to them but stays in the older leaves. This phenomenon is helpful in determining which nutrients a plant may be lacking.

Many plants engage in symbiosis with microorganisms. Two important types of these relationship are

1. with bacteria such as rhizobia, that carry out biological nitrogen fixation, in which atmospheric nitrogen (N_2) is converted into ammonium ($NH+4$); and

2. with mycorrhizal fungi, which through their association with the plant roots help to create a larger effective root surface area. Both of these mutualistic relationships enhance nutrient uptake.

Though nitrogen is plentiful in the Earth's atmosphere, relatively few plants harbour nitrogen-fixing bacteria, so most plants rely on nitrogen compounds present in the soil to support their growth. These can be supplied by mineralization of soil organic matter or added plant residues, nitrogen fixing bacteria, animal waste, through the breaking of triple bonded Nitrogen molecules by lightening strikes or through the application of fertilizers.

Functions of Nutrients

At least 17 elements are known to be essential nutrients for plants. In relatively large amounts, the soil supplies nitrogen, phosphorus, potassium, calcium, magnesium, and sulfur; these are often called the macronutrients. In relatively small amounts, the soil supplies iron, manganese, boron, molybdenum, copper, zinc, chlorine, and cobalt, the so-called micronutrients. Nutrients must be available not only in sufficient amounts but also in appropriate ratios.

Plant nutrition is a difficult subject to understand completely, partially because of the variation between different plants and even between different species or individuals of a given clone. Elements present at low levels may cause deficiency symptoms, and toxicity is possible at levels that are too high. Furthermore, deficiency of one element may present as symptoms of toxicity from another element, and vice versa. An abundance of one nutrient may cause a deficiency of another nutrient. For example, K^+ uptake can be influenced by the amount of $NH+4$ available.

Although nitrogen is plentiful in the Earth's atmosphere, relatively few plants engage in nitrogen fixation (conversion of atmospheric nitrogen to a biologically useful form). Most plants, therefore, require nitrogen compounds to be present in the soil in which they grow.

Carbon and oxygen are absorbed from the air while other nutrients are absorbed from the soil. Green plants obtain their carbohydrate supply from the carbon dioxide in the air by the process of photosynthesis. Each of these nutrients is used in a different place for a different essential function.

Macronutrients (Derived from air and Water)

Carbon

Carbon forms the backbone of most plant biomolecules, including proteins, starches and cellulose. Carbon is fixed through photosynthesis; this converts carbon dioxide from the air into carbohydrates which are used to store and transport energy within the plant.

Hydrogen

Hydrogen also is necessary for building sugars and building the plant. It is obtained almost entirely from water. Hydrogen ions are imperative for a proton gradient to help drive the electron transport chain in photosynthesis and for respiration.

Oxygen

Oxygen is a component of many organic and inorganic molecules within the plant, and is acquired in many forms. These include: O_2 and CO_2 (mainly from the air via leaves) and H_2O, NO–3, H_2PO–4 and SO2–4 (mainly from the soil water via roots). Plants produce oxygen gas (O_2) along with glucose during photosynthesis but then require O_2 to undergo aerobic cellular respiration and break down this glucose to produce ATP.

Macronutrients (Primary)

Nitrogen

Nitrogen is a major constituent of several of the most important plant substances. For example, nitrogen compounds comprise 40% to 50% of the dry matter of protoplasm, and it is a constituent of amino acids, the building blocks of proteins. It is also an essential constituent of chlorophyll. Nitrogen deficiency most often results in stunted growth, slow growth, and chlorosis. Nitrogen deficient plants will also exhibit a purple appearance on the stems, petioles and underside of leaves from an accumulation of anthocyanin pigments. Most of the nitrogen taken up by plants is from the soil in the forms of NO–3, although in acid environments such as boreal forests where nitrification is less likely to occur, ammonium NH+4 is more likely to be the dominating source of nitrogen. Amino acids and proteins can only be built from NH+4, so NO–3 must be reduced. In many agricultural settings, nitrogen is the limiting nutrient for rapid growth. Nitrogen is transported via the xylem from the roots to the leaf canopy as nitrate ions, or in an organic form, such as amino acids or amides. Nitrogen can also be transported in the phloem sap as amides, amino acids and ureides; it is therefore mobile within the plant, and the older leaves exhibit chlorosis and necrosis earlier than the younger leaves.

There is an abundant supply of nitrogen in the earth's atmosphere — N_2 gas comprises nearly 79% of air. However, N_2 is unavailable for use by most organisms because there is a triple bond between the two nitrogen atoms, making the molecule almost inert. In order for nitrogen to be used for growth it must be "fixed" (combined) in the form of ammonium (NH_4) or nitrate (NO_3) ions. The weathering of rocks releases these ions so slowly that it has a negligible effect on the availability of fixed nitrogen. Therefore, nitrogen is often the limiting factor for growth and biomass production in all environments where there is a suitable climate and availability of water to support life.

Nitrogen enters the plant largely through the roots. A "pool" of soluble nitrogen accumulates. Its composition within a species varies widely depending on several factors, including day length, time of day, night temperatures, nutrient deficiencies, and nutrient imbalance. Short day length promotes asparagine formation, whereas glutamine is produced under long day regimes. Darkness favors protein breakdown accompanied by high asparagine accumulation. Night temperature modifies the effects due to night length, and soluble nitrogen tends to accumulate owing to retarded synthesis and breakdown of proteins. Low night temperature conserves glutamine; high night temperature increases accumulation of asparagine because of breakdown. Deficiency of K accentuates differences between long- and short-day plants. The pool of soluble nitrogen is much smaller than in well-nourished plants when N and P are deficient since uptake of nitrate and further reduction and conversion of N to organic forms is restricted more than is protein synthesis. Deficiencies of Ca, K, and S affect the conversion of organic N to protein more than uptake and reduction. The size of the pool of soluble N is no guide *per se* to growth rate, but the size of the pool in relation to total N might be a useful ratio in this regard. Nitrogen availability in the rooting medium also affects the size and structure of tracheids formed in the long lateral roots of white spruce (Krasowski and Owens 1999).

Microorganisms have a central role in almost all aspects of nitrogen availability, and therefore for life support on earth. Some bacteria can convert N_2 into ammonia by the process termed *nitrogen fixation*; these bacteria are either free-living or form symbiotic associations with plants or other organisms (e.g., termites, protozoa), while other bacteria bring about transformations of ammonia to nitrate, and of nitrate to N_2 or other nitrogen gases. Many bacteria and fungi degrade organic matter, releasing fixed nitrogen for reuse by other organisms. All these processes contribute to the nitrogen cycle.

Phosphorus

Like nitrogen, phosphorus is involved with many vital plant processes. Within a plant, it is present mainly as a structural component of the nucleic acids, deoxyribonucleic nucleic acid (DNA) and ribose nucleic acid (RNA), and as a constituent of fatty phospholipids, of importance in membrane development and function. It is present in both organic and inorganic forms, both of which are readily translocated within the plant. All energy transfers in the cell are critically dependent on phosphorus. As with all living things, phosphorus is part of the Adenosine triphosphate (ATP), which is of immediate use in all processes that require energy with the cells. Phosphorus can also be used to modify the activity of various enzymes by phosphorylation, and is used for cell signaling. Phosphorus is concentrated at the most actively growing points of a plant and stored within seeds in anticipation of their germination. Phosphorus is most commonly found in the soil in the form of polyprotic phosphoric acid (H_3PO_4), but is taken up most readily in the form of H_2PO-4. Phosphorus is available to plants in limited quantities in most soils because it is released very slowly from insoluble phosphates and is rapidly fixed once again. Under most environmental conditions it is the element that limits growth because of this constriction and due to its high demand by plants and microorganisms. Plants can increase phosphorus uptake by a mutualism with mycorrhiza. A Phosphorus deficiency in plants is characterized by an intense green coloration or reddening in leaves due to lack of chlorophyll. If the plant is experiencing high phosphorus deficiencies the leaves may become denatured and show signs of death. Occasionally the leaves may appear purple from an accumulation of anthocyanin. Because phosphorus is a mobile nutrient, older leaves will show the first signs of deficiency.

On some soils, the phosphorus nutrition of some conifers, including the spruces, depends on the ability of mycorrhizae to take up, and make soil phosphorus available to the tree, hitherto unobtainable to the non-mycorrhizal root. Seedling white spruce, greenhouse-grown in sand testing negative for phosphorus, were very small and purple for many months until spontaneous mycorrhizal inoculation, the effect of which was manifested by a greening of foliage and the development of vigorous shoot growth.

Phosphorus deficiency can produce symptoms similar to those of nitrogen deficiency, but as noted by Russel: "Phosphate deficiency differs from nitrogen deficiency in being extremely difficult to diagnose, and crops can be suffering from extreme starvation without there being any obvious signs that lack of phosphate is the cause". Russell's observation applies to at least some coniferous seedlings, but Benzian found that although response to phosphorus in very acid forest tree nurseries in England was consistently high, no species (including Sitka spruce) showed any visible symptom of deficiency other than a slight lack of lustre. Phosphorus levels have to be exceedingly low before visible symptoms appear in such seedlings. In sand culture at o ppm phosphorus, white spruce seedlings were very small and tinted deep purple; at 0.62 ppm, only the smallest seedlings were deep purple; at 6.2 ppm, the seedlings were of good size and color.

It is useful to apply a high phosphorus content fertilizer, such as bone meal, to perennials to help with successful root formation.

Potassium

Unlike other major elements, potassium does not enter into the composition of any of the important plant constituents involved in metabolism, but it does occur in all parts of plants in substantial amounts. It seems to be of particular importance in leaves and at growing points. Potassium is outstanding among the nutrient elements for its mobility and solubility within plant tissues. Processes involving potassium include the formation of carbohydrates and proteins, the regulation of internal plant moisture, as a catalyst and condensing agent of complex substances, as an accelerator of enzyme action, and as contributor to photosynthesis, especially under low light intensity.

When soil-potassium levels are high, plants take up more potassium than needed for healthy growth. The term *luxury consumption* has been applied to this. When potassium is moderately deficient, the effects first appear in the older tissues, and from there progress towards the growing points. Acute deficiency severely affects growing points, and die-back commonly occurs. Symptoms of potassium deficiency in white spruce include: browning and death of needles (chlorosis); reduced growth in height and diameter; impaired retention of needles; and reduced needle length. A relationship between potassium nutrition and cold resistance has been found in several tree species, including two species of spruce.

Potassium regulates the opening and closing of the stomata by a potassium ion pump. Since stomata are important in water regulation, potassium regulates water loss from the leaves and increases drought tolerance. Potassium deficiency may cause necrosis or interveinal chlorosis. The potassium ion (K^+) is highly mobile and can aid in balancing the anion (negative) charges within the plant. Potassium helps in fruit coloration, shape and also increases its brix. Hence, quality fruits are produced in potassium-rich soils. Potassium serves as an activator of enzymes used in photosynthesis and respiration. Potassium is used to build cellulose and aids in photosynthesis by the formation of a chlorophyll precursor. Potassium deficiency may result in higher risk of pathogens, wilting, chlorosis, brown spotting, and higher chances of damage from frost and heat.

Macronutrients (Secondary and Tertiary)

Sulphur

Sulphur is a structural component of some amino acids and vitamins, and is essential in the manufacturing of chloroplasts. Sulphur is also found in the iron-sulphur complexes of the electron transport chains in photosynthesis. It is immobile and deficiency, therefore, affects younger tissues first. Symptoms of deficiency include yellowing of leaves and stunted growth.

Calcium

Calcium regulates transport of other nutrients into the plant and is also involved in the activation of certain plant enzymes. Calcium deficiency results in stunting. This nutrient is involved in photosynthesis and plant structure. Blossom end rot is also a result of inadequate calcium.

Calcium in plants occurs chiefly in the leaves, with lower concentrations in seeds, fruits, and roots. A major function is as a constituent of cell walls. When coupled with certain acidic compounds of the jelly-like pectins of the middle lamella, calcium forms an insoluble salt. It is also intimately involved in meristems, and is particularly important in root development, with roles in cell division, cell elongation, and the detoxification of hydrogen ions. Other functions attributed to calcium are; the neutralization of organic acids; inhibition of some potassium-activated ions; and a role in nitrogen absorption. A notable feature of calcium-deficient plants is a defective root system. Roots are usually affected before above-ground parts.

Magnesium

The outstanding role of magnesium in plant nutrition is as a constituent of the chlorophyll molecule. As a carrier, it is also involved in numerous enzyme reactions as an effective activator, in which it is closely associated with energy-supplying phosphorus compounds. Magnesium is very mobile in plants, and, like potassium, when deficient is translocated from older to younger tissues, so that signs of deficiency appear first on the oldest tissues and then spread progressively to younger tissues.

Micro-nutrients

Some elements are directly involved in plant metabolism (Arnon and Stout, 1939). However, this principle does not account for the so-called beneficial elements, whose presence, while not required, has clear positive effects on plant growth. Mineral elements that either stimulate growth but are not essential, or that are essential only for certain plant species, or under given conditions, are usually defined as beneficial elements.

Plants are able sufficiently to accumulate most trace elements. Some plants are sensitive indicators of the chemical environment in which they grow (Dunn 1991), and some plants have barrier mechanisms that exclude or limit the uptake of a particular element or ion species, e.g., alder twigs commonly accumulate molybdenum but not arsenic, whereas the reverse is true of spruce bark (Dunn 1991). Otherwise, a plant can integrate the geochemical signature of the soil mass permeated by its root system together with the contained groundwaters. Sampling is facilitated by the tendency of many elements to accumulate in tissues at the plant's extremities.

Iron

Iron is necessary for photosynthesis and is present as an enzyme cofactor in plants. Iron deficiency can result in interveinal chlorosis and necrosis. Iron is not a structural part of chlorophyll but very much essential for its synthesis. Copper deficiency can be responsible for promoting an iron deficiency.

Molybdenum

Molybdenum is a cofactor to enzymes important in building amino acids and is involved in nitrogen metabolism. Molybdenum is part of the nitrate reductase enzyme (needed for the reduction of nitrate) and the nitrogenase enzyme (required for biological nitrogen fixation).

Boron

Boron is found in the highly insoluble mineral, tourmaline. It is absorbed by plants in the form of the anion BO_3^{-3}. It is available to plants in moderately soluble mineral forms of Ca, Mg and Na borates and the highly soluble form of organic compounds. Concentration in soil must, in general, be below 5 ppm in the soil water solution, above that toxicity results. Its availability in soils ranges from 20 to 200 pounds per acre in the first eight inches, of which less than 5% is available. It is available to plants over a range of pH, from 5.0 to 7.5. It is mobile in the soil, hence, it is prone to leaching. Leaching removes substantial amounts of boron in sandy soil, but little in fine silt or clay soil. Boron's fixation to those minerals at high pH can render boron unavailable, while low pH frees the fixed boron, leaving it prone to leaching in wet climates. It precipitates with other minerals in the form of borax in which form it was first used over 400 years ago as a soil supplement. Decomposition of organic material causes boron to be deposited in the topmost soil layer; organic forms of boron are more soluble than their mineral form, hence are more available in the top few inches. When soil dries it can cause a precipitous drop in the availability of boron to plants as the plants cannot draw nutrients from that desiccated layer. Hence, boron deficiency diseases appear in dry weather.

Boron has many functions within a plant: it affects flowering and fruiting, pollen germination, cell division, and active salt absorption. The metabolism of amino acids and proteins, carbohydrates, calcium, and water are strongly affected by boron. Many of those listed functions may be embodied by its function in moving the highly polar sugars through cell membranes by reducing their polarity and hence the energy needed to pass the sugar. If sugar cannot pass to the fastest growing parts rapidly enough, those parts die. Boron is relatively immobile within a plant suggesting that the molecule is fixed to the points in the membrane where they facilitate sugar transport.

Boron is not relocatable in the plant via the phloem. It must be supplied to the growing parts via the xylem. Foliar sprays affect only those parts sprayed, which may be insufficient for the fastest growing parts, and is very temporary.

Boron is essential for the proper forming and strengthening of cell walls. Lack of boron results in short thick cells producing stunted fruiting bodies and roots. Calcium to boron ratio must be maintained in a narrow range for normal plant growth. For alfalfa, that calcium to boron ratio must be from 80:1 to 600:1. Boron deficiency appears at 800:1 and higher. For alfalfa, similar ratios exist for magnesium, copper, nitrogen and potassium. Boron levels within plants differ with plant species and range from 2.3 p.p.m for barley to 94.7 p.p.m for poppy . Lack of boron causes failure of calcium metabolism which produces hollow heart in beets and peanuts.

Inadequate amounts of boron affect many agricultural crops, legume forage crops most strongly. Of the micronutrients, boron deficiencies are second most common after zinc. Deficiencies of boron when soil is cropped are common and require the application of mineral supplement; one ton of alfalfa hay carries with it one ounce of boron, 100 bushels of peaches 4 ounces. Deficiency results in the death of the terminal growing points. Symptoms first appear as stunted growth, then to cellular changes, which leads to physical changes, and finally death of the plant.

Boron supplements derive from dry lake bed deposits such as those in Death Valley, USA, in the form of sodium tetraborate (borax), from which less soluble calcium borate is made. Foliar sprays are used on fruit crop trees in soils of high alkalinity. Boron is often applied to fields as a contaminant in other soil amendments but is not generally adequate to make up the rate of loss by cropping. The rates of application of borate to produce an adequate alfalfa crop range from 15 pounds per acre for a sandy-silt, acidic soil of low organic matter, to 60 pounds per acre for a soil with high organic matter, high cation exchange capacity and high pH.

Boron concentration in soil water solution higher than one ppm is toxic to most plants. Toxic concentrations within plants are 10 to 50 ppm for small grains and 200 ppm in boron-tolerant crops such as sugar beets, rutabaga, cucumbers, and conifers. Toxic soil conditions are generally limited to arid regions or can be caused by underground borax deposits in contact with water or volcanic gases dissolved in percolating water. Application rates should be limited to a few pounds per acre in a test plot to determine if boron is needed generally. Otherwise, testing for boron levels in plant material is required to determine remedies. Excess boron can be removed by irrigation and assisted by application of elemental sulfur to lower the pH and increase boron's solubility. Application of calcium will increase soil alkalinity, causing boron to fix on the mineral soil component and remove some fraction, thereby reducing boron toxicity.

Boron deficiencies must be detected by analysis of plant material to apply a correction before the obvious symptoms appear, after which it is too late to prevent crop loss. Strawberries deficient in boron will produce lumpy fruit; apricots will not blossom or, if they do, will not fruit or will drop their fruit depending on the level of boron deficit. Broadcast of boron supplements is effective and long term; a foliar spray is immediate but must be repeated.

Boron is an essential element for the health of animals which derive their boron from plant material.

Copper

Copper is important for photosynthesis. Symptoms for copper deficiency include chlorosis.It is involved in many enzyme processes; necessary for proper photosynthesis; involved in the manufacture of lignin (cell walls) and involved in grain production. It is also hard to find in some soil conditions.

Manganese

Manganese is necessary for photosynthesis, including the building of chloroplasts. Manganese deficiency may result in coloration abnormalities, such as discolored spots on the foliage.

Sodium

Sodium is involved in the regeneration of phosphoenolpyruvate in CAM and C4 plants. Sodium can potentially replace potassium's regulation of stomatal opening and closing.

Essentiality of sodium:

- Essential for C4 plants rather C3

- Substitution of K by Na: Plants can be classified into four groups:

    1. Group A—a high proportion of K can be replaced by Na and stimulate the growth, which cannot be achieved by the application of K

    2. Group B—specific growth responses to Na are observed but they are much less distinct

    3. Group C—Only minor substitution is possible and Na has no effect

    4. Group D—No substitution occurs

- Stimulate the growth—increase leaf area and stomata. Improves the water balance

- Na functions in metabolism

    1. C4 metabolism

    2. Impair the conversion of pyruvate to phosphoenol-pyruvate

    3. Reduce the photosystem II activity and ultrastructural changes in mesophyll chloroplast

- Replacing K functions

    1. Internal osmoticum

    2. Stomatal function

    3. Photosynthesis

    4. Counteraction in long distance transport

    5. Enzyme activation

- Improves the crop quality e.g. improves the taste of carrots by increasing sucrose

Zinc

Zinc is required in a large number of enzymes and plays an essential role in DNA transcription. A typical symptom of zinc deficiency is the stunted growth of leaves, commonly known as "little leaf" and is caused by the oxidative degradation of the growth hormone auxin.

Nickel

In higher plants, nickel is absorbed by plants in the form of Ni^{2+} ion. Nickel is essential for activation of urease, an enzyme involved with nitrogen metabolism that is required to process urea. Without nickel, toxic levels of urea accumulate, leading to the formation of necrotic lesions. In lower plants, nickel activates several enzymes involved in a variety of processes, and can substitute for zinc and iron as a cofactor in some enzymes.

Chlorine

Chlorine, as compounded chloride, is necessary for osmosis and ionic balance; it also plays a role in photosynthesis.

Cobalt

Cobalt has proven to be beneficial to at least some plants although it does not appear to be essential for most species. It has, however, been shown to be essential for nitrogen fixation by the nitrogen-fixing bacteria associated with legumes and other plants.

Aluminium

- Tea has a high tolerance for aluminum (Al) toxicity and the growth is stimulated by Al application. The possible reason is the prevention of Cu, Mn or P toxicity effects.

- There have been reports that Al may serve as a fungicide against certain types of root rot.

Silicon

Silicon is not considered an essential element for plant growth and development. It is always found in abundance in the environment and hence if needed it is available. It is found in the structures of plants and improves the health of plants.

In plants, silicon has been shown in experiments to strengthen cell walls, improve plant strength, health, and productivity. There have been studies showing evidence of silicon improving drought and frost resistance, decreasing lodging potential and boosting the plant's natural pest and disease fighting systems. Silicon has also been shown to improve plant vigor and physiology by improving root mass and density, and increasing above ground plant biomass and crop yields. Silicon is currently under consideration by the Association of American Plant Food Control Officials (AAPFCO) for elevation to the status of a "plant beneficial substance".

Higher plants differ characteristically in their capacity to take up silicon. Depending on their SiO_2 content they can be divided into three major groups:

- Wetland graminae-wetland rice, horsetail (10–15%)

- Dryland graminae-sugar cane, most of the cereal species and few dicotyledons species (1–3%)

- Most of dicotyledons especially legumes (<0.5%)

- The long distance transport of Si in plants is confined to the xylem. Its distribution within the shoot organ is therefore determined by transpiration rate in the organs

- The epidermal cell walls are impregnated with a film layer of silicon and effective barrier against water loss, cuticular transpiration rate in the organs.

Vanadium

Vanadium may be required by some plants, but at very low concentrations. It may also be substituting for molybdenum.

Selenium

Selenium is probably not essential for flowering plants, but it can be beneficial; it can stimulate plant growth, improve tolerance of oxidative stress, and increase resistance to pathogens and herbivory.

Selenium is, however, an essential mineral element for animal (including human) nutrition and selenium deficiencies are known to occur when food or animal feed is grown on selenium-deficient soils. The use of inorganic selenium fertilizers can increase selenium concentrations in edible crops and animal diets thereby improving animal health.

Nutrient Deficiency

The effect of a nutrient deficiency can vary from a subtle depression of growth rate to obvious stunting, deformity, discoloration, distress, and even death. Visual symptoms distinctive enough to be useful in identifying a deficiency are rare. Most deficiencies are multiple and moderate. However, while a deficiency is seldom that of a single nutrient, nitrogen is commonly the nutrient in shortest supply.

Chlorosis of foliage is not always due to mineral nutrient deficiency. Solarization can produce superficially similar effects, though mineral deficiency tends to cause premature defoliation, whereas solarization does not, nor does solarization depress nitrogen concentration.

Nutrient Status of Plants

Nutrient status (mineral nutrient and trace element composition, also called ionome and nutrient profile) of plants are commonly portrayed by tissue elementary analysis. Interpretation of the results of such studies, however, has been controversial. During the last decades the nearly two-century-old "law of minimum" or "Liebig's law" (that states that plant growth is controlled not by the total amount of resources available, but by the scarcest resource) has been replaced by several mathematical approaches that use different models in order to take the interactions between the individual nutrients into account. The latest developments in this field are based on the fact that the nutrient elements (and compounds) do not act independently from each other; Baxter, 2015, because there may be direct chemical interactions between them or they may influence each other's uptake, translocation, and biological action via a number of mechanisms as exemplified for the case of ammonia.

Plant Nutrition in Agricultural Systems

Hydroponics

Hydroponics is a method for growing plants in a water-nutrient solution without the use of nutrient-rich soil. It allows researchers and home gardeners to grow their plants in a controlled environment. The most common solution is the Hoagland solution, developed by D. R. Hoagland in 1933. The solution consists of all the essential nutrients in the correct proportions necessary for most plant growth. An aerator is used to prevent an anoxic event or hypoxia. Hypoxia can affect nutrient uptake of a plant because, without oxygen present, respiration becomes inhibited within the root cells. The nutrient film technique is a hydroponic technique in which the roots are not fully submerged. This allows for adequate aeration of the roots, while a "film" thin layer of nutrient-rich water is pumped through the system to provide nutrients and water to the plant.

Water Potential

Water potential is the potential energy of water per unit volume relative to pure water in reference conditions. Water potential quantifies the tendency of water to move from one area to another due to osmosis, gravity, mechanical pressure, or matrix effects such as capillary action (which is caused by surface tension). The concept of water potential has proved useful in understanding and computing water movement within plants, animals, and soil. Water potential is typically expressed in potential energy per unit volume and very often is represented by the Greek letter ψ.

Water potential integrates a variety of different potential drivers of water movement, which may operate in the same or different directions. Within complex biological systems, many potential factors may be operating simultaneously. For example, the addition of solutes lowers the potential (negative vector), while an increase in pressure increases the potential (positive vector). If flow is not restricted, water will move from an area of higher water potential to an area that is lower potential. A common example is water with a dissolved salt, such as sea water or the fluid in a living cell. These solutions have negative water potential, relative to the pure water reference. With no restriction on flow, water will move from the locus of greater potential (pure water) to the locus of lesser (the solution); flow proceeds until the difference in potential is equalized or balanced by another water potential factor, such as pressure or elevation.

Components of Water Potential

Many different factors may affect the total water potential, and the sum of these potentials determines the overall water potential and the direction of water flow:

$$\Psi = \Psi_0 + \Psi_\pi + \Psi_p + \Psi_s + \Psi_v + \Psi_m$$

where:

- $\emptyset_0$ is the reference correction,

- Ψ_π is the solute or osmotic potential,

- Ψ_p is the pressure component,

- Ψ_s is the gravimetric component,

-     is the potential due to humidity, and

- Ψ_m is the potential due to matrix effects (e.g., fluid cohesion and surface tension.)

All of these factors are quantified as potential energies per unit volume, and different subsets of these terms may be used for particular applications (e.g., plants or soils). Different conditions are also defined as reference depending on the application: for example, in soils, the reference condition is typically defined as pure water at the soil surface.

Pressure Potential

Pressure potential is based on mechanical pressure, and is an important component of the total water potential within plant cells. Pressure potential increases as water enters a cell. As water passes through the cell wall and cell membrane, it increases the total amount of water present inside the cell, which exerts an outward pressure that is opposed by the structural rigidity of the cell wall. By creating this pressure, the plant can maintain turgor, which allows the plant to keep its rigidity. Without turgor, plants will lose structure and wilt.

The pressure potential in a plant cell is usually positive. In plasmolysed cells, pressure potential is almost zero. Negative pressure potentials occur when water is pulled through an open system such as a plant xylem vessel. Withstanding negative pressure potentials (frequently called *tension*) is an important adaptation of xylem. This tension can be measured empirically using the Pressure bomb.

Osmotic Potential (Solute Potential)

Pure water is usually defined as having an osmotic potential (Ψ_π) of zero, and in this case, solute potential can never be positive. The relationship of solute concentration (in molarity) to solute potential is given by the van 't Hoff equation:

$$\Psi_\pi = -MiRT$$

where M is the concentration in molarity of the solute, i is the van 't Hoff factor, the ratio of amount of particles in solution to amount of formula units dissolved, R is the ideal gas constant, and T is the absolute temperature.

For example, when a solute is dissolved in water, water molecules are less likely to diffuse away via osmosis than when there is no solute. A solution will have a lower and hence more negative water potential than that of pure water. Furthermore, the more solute molecules present, the more negative the solute potential is.

Osmotic potential has important implications for many living organisms. If a living cell is surrounded by a more concentrated solution, the cell will tend to lose water to the more negative water potential (Ψ_w) of the surrounding environment. This can be the case for marine organisms living in sea water and halophytic plants growing in saline environments. In the case of a plant cell, the flow of water out of the cell may eventually cause the plasma membrane to pull away from the cell wall, leading to plasmolysis. Most plants, however, have the ability to increase solute inside the cell to drive the flow of water into the cell and maintain turgor.

This effect can be used to power an osmotic power plant.

A soil solution also experiences osmotic potential. The osmotic potential is made possible due to the presence of both inorganic and organic solutes in the soil solution. As water molecules increasingly clump around solute ions or molecules, the freedom of movement, and thus the potential energy, of the water is lowered. As the concentration of solutes is increased, the osmotic potential of the soil solution is reduced. Since water has a tendency to move toward lower energy levels, water will want to travel toward the zone of higher solute concentrations. Although, liquid water will only move in response to such differences in osmotic potential if a semipermeable membrane exists between the zones of high and low osmotic potential. A semipermeable membrane is necessary because it allows water through its membrane while preventing solutes from moving through its membrane. If no membrane is present, movement of the solute, rather than of the water, largely equalizes concentrations.

Since regions of soil are usually not divided by a semipermeable membrane, the osmotic potential typically has a negligible influence on the mass movement of water in soils. On the other hand, osmotic potential has an extreme influence on the rate of water uptake by plants. If soils are high in soluble salts, the osmotic potential is likely to be lower in the soil solution than in the plant root cells. In such cases, the soil solution would severely restrict the rate of water uptake by plants. In salty soils, the osmotic potential of soil water may be so low that the cells in young seedlings start to collapse (plasmolyze).

Matrix potential (Matric potential)

When water is in contact with solid particles (e.g., clay or sand particles within soil), adhesive intermolecular forces between the water and the solid can be large and important. The forces between the water molecules and the solid particles in combination with attraction among water molecules promote surface tension and the formation of menisci within the solid matrix. Force is then required to break these menisci. The magnitude of matrix potential depends on the distances between solid particles—the width of the menisci (also capillary action and differing Pa at ends of capillary)—and the chemical composition of the solid matrix (meniscus, macroscopic motion due to ionic attraction).

In many cases, matrix potential can be relatively large in comparison to the other components of water potential discussed above. Matrix potential markedly reduces the energy state of water near particle surfaces. Although water movement due to matrix potential may be slow, it is still extremely important in supplying water to plant roots and in engineering applications. The matrix potential is always negative because the water attracted by the soil matrix has an energy state lower than that of pure water. Matrix potential only occurs in unsaturated soil above the water table. If the matrix potential approaches a value of zero, nearly all soil pores are completely filled with water, i.e. fully saturated and at maximum retentive capacity. The matrix potential can vary considerably among soils. In the case that water drains into less-moist soil zones of similar porosity, the matrix potential is generally in the range of −10 to −30 kPa.

It is worth noting that matrix potentials are very important for plant water relations. Strong (very negative) matrix potentials bind water to soil particles within very dry soils. Plants then create even more negative matrix potentials within tiny pores in the cell walls of their leaves to extract water from the soil and allow physiological activity to continue through dry periods. Germinating seeds have a very negative matrix potentials, creating water uptake in even somewhat dry soils and hydrates the dry seed.

Empirical Examples

Soil-plant-air Continuum

At a potential of 0 kPa, soil is in a state of saturation. At saturation, all soil pores are filled with water, and water typically drains from large pores by gravity. At a potential of −33 kPa, or −1/3 bar, (−10 kPa for sand), soil is at field capacity. Typically, at field capacity, air is in the macropores and water in micropores. Field capacity is viewed as the optimal condition for plant growth and microbial activity. At a potential of −1500 kPa, soil is at its permanent wilting point, meaning that soil water is held by solid particles as a "water film" that is retained too tightly to be taken up by plants.

In contrast, atmospheric water potentials are much more negative—a typical value for dry air is −100 MPa, though this value depends on the temperature and the humidity. Root water potential must be more negative than the soil, and the stem water potential and intermediate lower value than the roots but higher than the leaf water potential, to create a passive flow of water from the soil to the roots, up the stem, to the leaves and then into the atmosphere.

Measurement Techniques

A tensiometer, electrical resistance gypsum block, neutron probes, or time-domain reflectometry (TDR) can be used to determine soil water potential. Tensiometers are limited to 0 to −85 kPa, electrical resistance blocks is limited to −90 to −1500 kPa, neutron probes is limited to 0 to −1500 kPa, and TDR is limited to 0 to −10,000 kPa. A scale can be used to estimate water weight (percentage composition) if special equipment is not on hand.

Transpiration

Overview of transpiration:

1. Water is passively transported into the roots and then into the xylem.

2. The forces of cohesion and adhesion cause the water molecules to form a column in the xylem.

3. Water moves from the xylem into the mesophyll cells, evaporates from their surfaces and leaves the plant by diffusion through the stomata

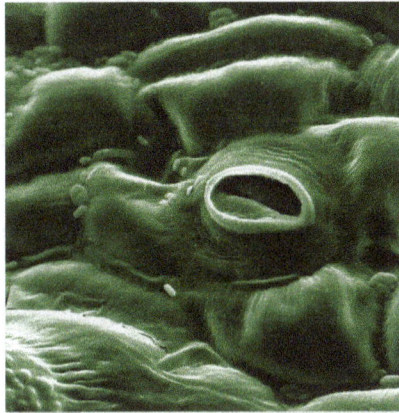

Stoma in a tomato leaf shown via colorized scanning electron microscope

The clouds in this image of the Amazon Rainforest are a result of transpiration.

Transpiration is the process of water movement through a plant and its evaporation from aerial parts, such as leaves, stems and flowers. Water is necessary for plants but only a small amount of water taken up by the roots is used for growth and metabolism. The remaining 97–99.5% is lost by transpiration and guttation. Leaf surfaces are dotted with pores called stomata, and in most plants they are more numerous on the undersides of the foliage. The stomata are bordered by guard cells and their stomatal accessory cells (together known as stomatal complex) that open and close the pore. Transpiration occurs through the stomatal apertures, and can be thought of as a necessary "cost" associated with the opening of the stomata to allow the diffusion of carbon dioxide gas from the air for photosynthesis. Transpiration also cools plants, changes osmotic pressure of cells, and enables mass flow of mineral nutrients and water from roots to shoots. Two major factors influence the rate of water flow from the soil to the roots: the hydraulic conductivity of the soil and the magnitude of the pressure gradient through the soil. Both of these factors influence the rate of bulk flow of water moving from the roots to the stomatal pores in the leaves via the xylem.

Mass flow of liquid water from the roots to the leaves is driven in part by capillary action, but primarily driven by water potential differences. If the water potential in the ambient air is lower than the water potential in the leaf airspace of the stomatal pore, water vapor will travel down the gradient and move from the leaf airspace to the atmosphere. This movement lowers the water potential in the leaf airspace and causes evaporation from the mesophyll cell wall menisci of liquid water. This evaporation increases the tension on the menisci surface and increases its radius. Because of the cohesive properties of water, the tension travels through the leaf cells to the leaf and stem xylem where a momentary negative pressure is created as water is pulled up the xylem from the roots. In taller plants and trees, the force of gravity can only be overcome by the decrease in hydrostatic (water) pressure in the upper parts of the plants due to the diffusion of water out of stomata into the atmosphere. Water is absorbed at the roots by osmosis, and any dissolved mineral nutrients travel with it through the xylem.

The Cohesion-tension theory explains how leaves pull water through the xylem. Water molecules stick together, or exhibit cohesion. As a water molecule evaporates from the surface of the leaf, it pulls on the adjacent water molecule, creating a continuous flow of water through the plant.

Regulation

Plants regulate the rate of transpiration by controlling the size of the stomatal apertures. The rate of transpiration is also influenced by the evaporative demand of the atmosphere surrounding the leaf such as boundary layer conductance, humidity, temperature, wind and incident sunlight. Soil water supply and soil temperature can influence stomatal opening, and thus transpiration rate. The amount of water lost by a plant also depends on its size and the amount of water absorbed at the roots. Transpiration accounts for most of the water loss by a plant by the leaves and young stems. Transpiration serves to evaporatively cool plants, as the evaporating water carries away heat energy due to its large latent heat of vaporization of 2260 kJ per litre.

Feature	Effect on transpiration
Number of leaves	More leaves (or spines, or other photosynthesizing organs) means a bigger surface area and more stomata for gaseous exchange. This will result in greater water loss.
Number of stomata	More stomata will provide more pores for transpiration.
Size of the leaf	A leaf with a bigger surface area will transpire faster than a leaf with a smaller surface area.
Presence of plant cuticle	A waxy cuticle is relatively impermeable to water and water vapour and reduces evaporation from the plant surface except via the stomata. A reflective cuticle will reduce solar heating and temperature rise of the leaf, helping to reduce the rate of evaporation. Tiny hair-like structures called trichomes on the surface of leaves also can inhibit water loss by creating a high humidity environment at the surface of leaves. These are some examples of the adaptations of plants for conservation of water that may be found on many xerophytes.
Light supply	The rate of transpiration is controlled by stomatal aperture, and these small pores open especially for photosynthesis. While there are exceptions to this (such as night or "CAM photosynthesis"), in general a light supply will encourage open stomata.

Temperature	Temperature affects the rate in two ways: 1) An increased rate of evaporation due to a temperature rise will hasten the loss of water. 2) Decreased *relative* humidity outside the leaf will increase the water potential gradient.
Relative humidity	Drier surroundings gives a steeper water potential gradient, and so increases the rate of transpiration.
Wind	In still air, water lost due to transpiration can accumulate in the form of vapor close to the leaf surface. This will reduce the rate of water loss, as the water potential gradient from inside to outside of the leaf is then slightly less. Wind blows away much of this water vapor near the leaf surface, making the potential gradient steeper and speeding up the diffusion of water molecules into the surrounding air. Even in wind, though, there may be some accumulation of water vapor in a thin boundary layer of slower moving air next to the leaf surface. The stronger the wind, the thinner this layer will tend to be, and the steeper the water potential gradient.
Water supply	Water stress caused by restricted water supply from the soil may result in stomatal closure and reduce the rates of transpiration.

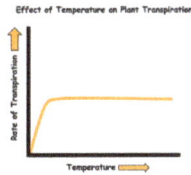

The effect of temperature on the transpiration rate of plants.

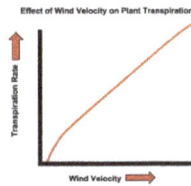

The effect of wind velocity on the transpiration rate of plants.

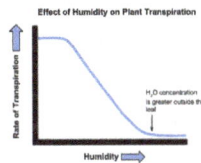

The effect of humidity on the transpiration rate of plants.

Some xerophytes will reduce the surface of their leaves during water deficiencies (left).
If temperatures are cool enough and water levels are adequate the leaves expand again (right).

During a growing season, a leaf will transpire many times more water than its own weight. An acre of corn gives off about 3,000–4,000 gallons (11,400–15,100 liters) of water each day, and a large oak tree can transpire 40,000 gallons (151,000 liters) per year. The transpiration ratio is the ratio of the mass of water transpired to the mass of dry matter produced; the transpiration ratio of crops tends to fall between 200 and 1000 (*i.e.*, crop plants transpire 200 to 1000 kg of water for every kg of dry matter produced).

Transpiration rates of plants can be measured by a number of techniques, including potometers, lysimeters, porometers, photosynthesis systems and thermometric sap flow sensors. Isotope measurements indicate transpiration is the larger component of evapotranspiration. Recent evidence from a global study of water stable isotopes shows that transpired water is isotopically different from groundwater and streams. This suggests that soil water is not as well mixed as widely assumed.

Desert plants have specially adapted structures, such as thick cuticles, reduced leaf areas, sunken stomata and hairs to reduce transpiration and conserve water. Many cacti conduct photosynthesis in succulent stems, rather than leaves, so the surface area of the shoot is very low. Many desert plants have a special type of photosynthesis, termed crassulacean acid metabolism or CAM photosynthesis, in which the stomata are closed during the day and open at night when transpiration will be lower.

References

- Allen V. Barker; D. J. Pilbeam (2007). Handbook of plant nutrition. CRC Press. ISBN 978-0-8247-5904-9. Retrieved 17 August 2010.

- Marschner, Petra, ed. (2012). Marschner's mineral nutrition of higher plants (3rd ed.). Amsterdam: Elsevier/Academic Press. ISBN 9780123849052.

- Norman P. A. Huner; William Hopkins. "3 & 4". Introduction to Plant Physiology 4th Edition. John Wiley & Sons, Inc. ISBN 978-0-470-24766-2.

- Barker, AV; Pilbeam, DJ (2015). Handbook of Plant Nutrition. (2nd ed.). CRC Press. ISBN 9781439881972. Retrieved 5 June 2016.

- Jones, Hamlyn G. (2013-12-12). Plants and Microclimate: A Quantitative Approach to Environmental Plant Physiology. Cambridge University Press. p. 93. ISBN 9781107511637.

- Taiz, Lincoln (2015). Plant Physiology and Development. Sunderland, MA: Sinauer Associates, Inc. p. 101. ISBN 978-1-60535-255-8.

- Graham, Linda E. (2006). Plant Biology. Upper Saddle River, NJ 07458: Pearson Education, Inc. pp. 200–202. ISBN 0-13-146906-1.

- Martin, J.; Leonard, W.; Stamp, D. (1976), Principles of Field Crop Production (3rd ed.), New York: Macmillan Publishing Co., ISBN 0-02-376720-0.

- White, Philip J. (2016). "Selenium accumulation by plants". Annals of Botany. 117: 217–235. doi:10.1093/aob/mcv180. Retrieved 5 June 2016.

- Bowen, Gabriel (2015-09-03). "Hydrology: The diversified economics of soil water". Nature. 525 (7567): 43–44. doi:10.1038/525043a. ISSN 0028-0836.

Permissions

Index